Springer Topics in Signal Processing

Volume 9

More information about this series at http://www.springer.com/series/8109

Daniel P. Jarrett · Emanuël A.P. Habets
Patrick A. Naylor

Theory and Applications of Spherical Microphone Array Processing

Springer

Daniel P. Jarrett
Kilburn & Strode LLP
London
UK

Emanuël A.P. Habets
International Audio Laboratories Erlangen
Erlangen
Germany

Patrick A. Naylor
Department of Electrical and Electronic Engineering
Imperial College London
London
UK

ISSN 1866-2609 ISSN 1866-2617 (electronic)
Springer Topics in Signal Processing
ISBN 978-3-319-42209-1 ISBN 978-3-319-42211-4 (eBook)
DOI 10.1007/978-3-319-42211-4

Library of Congress Control Number: 2016944322

Printed on acid-free paper

This Springer imprint is published by Springer Nature
The registered company is Springer International Publishing AG Switzerland

Preface

The topic of spherical microphone array signal processing has been gaining importance since the publications of Meyer and Elko around 2002, and fuelled by many others since.

Sound is unavoidably influenced by the space in which it is rendered, as we all know from personal experience, and the capability of microphone arrays to capture the spatial information is both fascinating and intriguing. The English physicist Charles Wheatstone is credited with the first use of the term 'microphone'. However, it was not until the carbon microphone, invented by David Hughes and demonstrated in 1877, that the concept of capturing sound as an electrical signal became established. The invention by Gerhard Sessler and Jim West of the electret microphone in 1962, and further developments of condenser microphone technology in particular, led to a significant improvement in quality and reliability.

These early microphones were principally targeting the capture of acoustic signals in a close-talking mode, less than around 10 cm from the talker's or singer's lips. Would their inventors have considered that the spatial information associated with the sound could be useful, exploited to localize sources of sound, discriminate desired sounds from interferences, or even infer the geometry of an acoustic space and navigate within it? We could only guess but certainly the potential to achieve these goals has been always present. The catalyst for more recent developments has been the happy marriage of high quality, synchronized multichannel analogue-to-digital conversion with powerful digital signal processing hardware and software, facilitating arrays with elements numbering from a handful to hundreds, or even thousands. Given the availability of numerous sensors, many alternative geometries can be considered, the spherical geometry being one such with considerable merits. It is the algorithms to process the signals from these numerous microphones that are the key focus of this book.

We offer the reader a view of the theoretical aspects of microphone array signal processing for spherical geometries and some examples of applications of the ensuing algorithms. Our intention is to present the methods in a general form allowing the ideas to be further developed. It is a well known feeling that digging deeper into a subject only serves to reveal greater depths and further potential. We hope, nevertheless, that this book will at the same time provide satisfaction to the mind of the curious reader but also serve to equip the researchers of the future to develop and exploit the great potential of spherical microphone arrays and their associated signal processing.

We gratefully acknowledge the contributions of Sebastian Braun, Maja Taseska, Oliver Thiergart and Mark Thomas to the work presented in this book. We would also like to express our gratitude to Hamza Javed and Maja Taseska for their attentive reading of our drafts, and to Sira Gonzalez and Felicia Lim for providing helpful feedback and suggestions throughout the writing process.

London Daniel P. Jarrett
Erlangen Emanuël A.P. Habets
London Patrick A. Naylor
April 2016

Contents

Abbreviations

AIR	Acoustic impulse response
ATF	Acoustic transfer function
BRIR	Binaural room impulse response
CASA	Computational auditory scene analysis
CV	Coefficient of variation
DI	Directivity index
DirAC	Directional Audio Coding
DOA	Direction of arrival
DSPP	Desired speech presence probability
GSC	Generalized sidelobe canceller
HARPEX	High angular resolution planewave expansion
HRIR	Head-related impulse response
HRTF	Head-related transfer function
ILD	Interaural level difference
iSINR	Input signal-to-incoherent-noise ratio
ITD	Interaural time difference
LCMP	Linearly constrained minimum power
LCMV	Linearly constrained minimum variance
MPDR	Minimum power distortionless response
MSE	Mean square error
MVDR	Minimum variance distortionless response
NRF	Noise reduction factor
OLA	Overlap-add
PDF	Probability distribution function
PSD	Power spectral density
PWD	Plane-wave decomposition
RTF	Relative transfer function
SDR	Signal-to-diffuse ratio
SHD	Spherical harmonic domain
SHE	Spherical harmonic expansion

SHT	Spherical harmonic transform
SMIR	Spherical Microphone array Impulse Response (method)
SNR	Signal-to-noise ratio
SPP	Speech presence probability
SRA	Statistical room acoustics
SRMR	Speech-to-reverberation modulation energy ratio
SRP	Steered response power
STFT	Short-time Fourier transform
WNG	White noise gain
WOLA	Weighted overlap-add

Operators

$x * y$	Linear convolution operator
$[\cdot]^*$	Complex conjugate
$[\cdot]^{\mathrm{T}}$	Vector/matrix transpose
$[\cdot]^{-1}$	Matrix inverse
$[\cdot]^{\mathrm{H}}$	Hermitian transpose
$(\cdot)!$	Factorial
$\lvert\cdot\rvert$	Absolute value
$\lvert\lvert\cdot\rvert\rvert$	Matrix/vector 2-norm
$\mathrm{diag}\{\cdot\}$	Diagonal operator
$\mathrm{E}\{\cdot\}$	Mathematical expectation
$\Re\{\cdot\}$	Real part of a complex number
$\Im\{\cdot\}$	Imaginary part of a complex number
$\mathrm{tr}\{\cdot\}$	Matrix trace
$f'(x)$	Derivative of f with respect to x

Symbols and Variables

b_l	Mode strength of order l
β_{ab}	Room boundary reflection coefficient, $a \in \{x, y, z\}, b \in \{1, 2\}$
c	Speed of sound
$\mathbf{d}$	Relative transfer function vector
$\delta(\cdot)$	Dirac delta function
$\delta_{.}$	Kronecker delta
ε	Angular error
f	Frequency
$f(\cdot)$	Probability distribution function
γ	Spatial coherence
Γ	Signal-to-diffuse ratio
g	Time domain free-space Green's function
G	Frequency domain free-space Green's function
G_{N}	Frequency domain Neumann Green's function
$\mathfrak{g}_{q,lm}$	Quadrature weight for microphone q, order l and degree m
h	Acoustic impulse response (time domain)
H	Acoustic transfer function (frequency domain)
$h_l^{(1)}$	Spherical Hankel function of the first kind and of order l
$h_l^{(2)}$	Spherical Hankel function of the second kind and of order l
i	Complex number, $i^2 = -1$
$\mathcal{I}$	Intensity vector
$\mathbf{I}$	Pseudointensity vector
$\mathbf{I}_{N \times N}$	$N \times N$ identity matrix
j_l	Spherical Bessel function of order l
k	Wavenumber (continuous)
λ	Lagrange multiplier
l	Order
ℓ	Time frame index
L	Array order (maximum spherical harmonic order)
m	Degree

$\mathcal{M}_{\mathrm{ref}}$	Reference microphone (at centre of sphere)
μ	Tradeoff parameter
n	Time (discrete)
ν	Frequency index
ω	Angular frequency
$\Omega = (\theta, \phi)$	Spherical angular coordinates (inclination θ, azimuth ϕ)
Ω_0	Direction of arrival
Ω_{u}	Look direction
Ψ	Diffuseness
$\mathbf{\Phi}$	Covariance matrix
$\mathcal{P}_l$	Legendre polynomial of order l
$\mathcal{P}_{lm}$	Associated Legendre function of order l and degree m
q	Microphone index
Q	Number of microphones
$\mathbf{r}$	Receiver position vector
$\mathbf{r}_{\mathrm{s}}$	Source position vector
R_{lm}	Real spherical harmonic of order l and degree m
ρ_0	Density of air
t	Time (continuous)
T_{60}	Reverberation time
$\mathbf{u}$	Unit vector pointing towards acoustic source
$\hat{\mathbf{u}}$	Estimate of $\mathbf{u}$
$\mathbf{v}$	Particle velocity vector
$\mathbf{w}$	Filter weights vector
Y_{lm}	Complex spherical harmonic of order l and degree m
Z	Beamformer output signal

Chapter 1
Introduction

1.1 Background and Context

The motivation behind this book lies in the rapidly growing interest in spherical microphone arrays over the last decade. Important applications for these arrays include human-human and human-machine speech communication systems and spatial sound recording. While human-human speech communication systems have a long history, speech also plays an ever-growing part in human-machine communication. Indeed, while speech-based interfaces were once confined to the realms of science fiction, they are now becoming an increasingly popular way of interacting with devices such as smartphones, desktop and tablet computers, robots or televisions. This trend has been fuelled by advances in speech recognition technology, as well as the explosion in available computing power, particularly on mobile devices. With the widespread availability of 3D sound cinema systems and virtual reality gear with 3D binaural sound reproduction, the need to capture spatial sound is rapidly growing. Spherical microphone arrays are particularly suitable for capturing all three dimensions of the sound field, including both ambient sounds and sounds from particular directions.

The field of acoustic signal processing seeks to solve a number of problems relating to these systems, which can broadly be divided into three categories: acoustic parameter estimation, acoustic signal enhancement, and spatial audio recording. Acoustic parameter estimation, addressed in Chap. 5, involves the estimation of parameters such as the location or direction of arrival (DOA) of one or more acoustic sources [20, 27, 30, 34, 51–53], the signal-to-diffuse energy ratio or diffuseness of the sound field at a particular position [31, 32, 43, 54, 55], the number of sources present in a sound field [53, 56, 57], or the reverberation time of an acoustic environment [14, 39, 40, 46, 49, 58].

Portions of this chapter were first published in [25], and are reproduced here with the author's permission.

D. Jarrett et al., *Theory and Applications of Spherical Microphone Array Processing*, Springer Topics in Signal Processing 9,
DOI 10.1007/978-3-319-42211-4_1

In the aforementioned applications, the signal to be acquired originates from a *distant* source, located at some significant distance from the microphone(s). While in some applications, such as teleconferencing systems, a microphone located close to the source may be available, this is not always a practical option. As a result, the acquired signal is corrupted by the surrounding environment. One major cause for this degradation is the presence of noise, where by *noise* we mean any acoustic signal which is undesired, such as interfering speech signals or background noise [6, 9]. The other is the presence of reflectors and obstacles to the propagation of sound waves, in particular room boundaries (walls, floors and ceiling), which cause *reverberation* [35, 42]. As the distance between the source and microphone(s) increases, the degradative effects of noise and reverberation become increasingly significant.

In the case of speech signals, these effects not only degrade the quality of the acquired signal, but in some cases also its intelligibility, making communication difficult or even impossible [3]. The cognitive effort required to understand highly noisy and reverberant speech can also contribute to listener fatigue. Acoustic signal enhancement or speech enhancement techniques (considered in Chaps. 6–9) seek to mitigate these effects, and extract the desired signal. The main problems of interest within this field are noise reduction [4, 6, 22, 26, 28], echo cancellation [7, 33, 47] and dereverberation [11, 18, 21, 23, 24, 29, 37, 38, 42]. Although the release of the first speakerphone dates back to 1954 [15], these remain open problems and areas of active research.

1.2 Microphone Array Signal Processing

Acoustic signal processing problems are commonly approached with microphone arrays [5, 10, 17], which is an arrangement of microphones in a specific configuration, thereby taking advantage of the spatial properties of the sound field (or *spatial diversity*) in order to improve performance. Owing to the similarity of the problems involved, many microphone array processing techniques are based on narrowband antenna array processing techniques [12]; however, microphone array processing faces its own unique challenges [5]. These include the broadband nature of speech (which covers several octaves), the non-stationarity of speech, and the fact that the desired and noise signals often have very similar spectral characteristics [5]. In addition, the placement and number of microphones is restricted, primarily by cost, aesthetics and available space. Considerations of space limit both the inter-microphone spacing and total microphone array size, and are of particular importance for portable devices, such as hearing aids [13].

A typical application scenario in microphone array signal processing is illustrated in Fig. 1.1. A microphone array captures a mixture of signals with different spatial characteristics, some of which may be desired, and others undesired. Acoustic parameter estimation algorithms seek to accurately estimate the parameters of interest even in the presence of undesired signals that may adversely affect the estimation

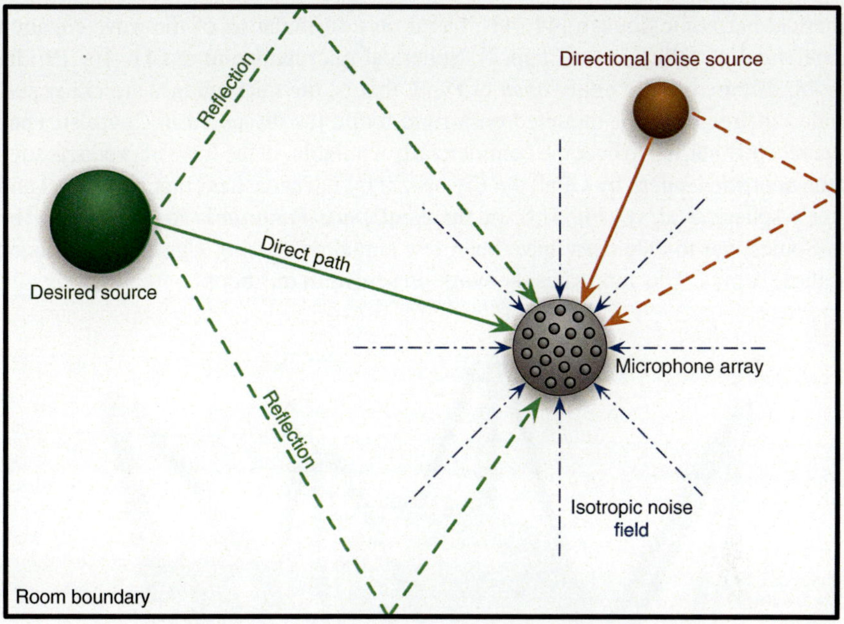

Fig. 1.1 Schematic illustration of a typical application scenario in microphone array signal processing. A microphone array captures a mixture of signals with different spatial characteristics in a reverberant environment

process. Acoustic signal enhancement algorithms aim to extract only the desired signals from the received mixture.

The spatial characteristics of the various captured signals are typically modeled based on their *spatial coherence*. The microphone signals are corrupted by sensor noise, which is spatially *incoherent* (or *spatially white*), that is, the sensor noise signals at each microphone are mutually uncorrelated. The desired signals, originating from one or more desired sources, as well as any directional noise signals, originating from interfering speakers or air-conditioning units, for example, are spatially *coherent*. Finally, *partially coherent* signals can be observed in spherically or cylindrically *isotropic* (or diffuse) sound fields, which can be used to model babble noise or reverberation. The desired signal is normally chosen as either the anechoic signal arriving from the desired source via the *direct path*, or the reverberant signal arriving via the direct path and a number of reflected paths.

In theory, any microphone array configuration is possible; in practice, most microphone arrays are linear or planar, and the microphones respectively lie on a straight line or a flat, two-dimensional surface. Real sound fields are three-dimensional, however, and can only be fully analyzed with a three-dimensional array. The spherical configuration is convenient due to its symmetry giving equal performance in all directions. In addition, the captured sound field can be efficiently described in the spherical harmonic domain [41, 44], based on a formulation of the wave equation in spherical coordinates (in Chap. 2). Spherical microphone arrays [1, 16, 19, 36, 45, 48, 50] are usually either *open* or *rigid*, that is, the microphones are either suspended in free space or mounted on a rigid baffle (as discussed in Chap. 3). They have recently started to become commercially available, in the form of products such as the *acoustic camera* by GFal, the *Eigenmike* by mh acoustics (Fig. 1.2), Brüel and Kjær's spherical array (Fig. 1.3), or the *RealSpace Panoramic Audio Camera* by VisiSonics, yet to date there have been few signal processing algorithms designed for these arrays. This motivates the work presented in this book.

Fig. 1.2 The em32 Eigenmike spherical microphone array. This rigid array of radius 4.2 cm is comprised of 32 omnidirectional microphones. Copyright © Emanuël Habets. Used with permission

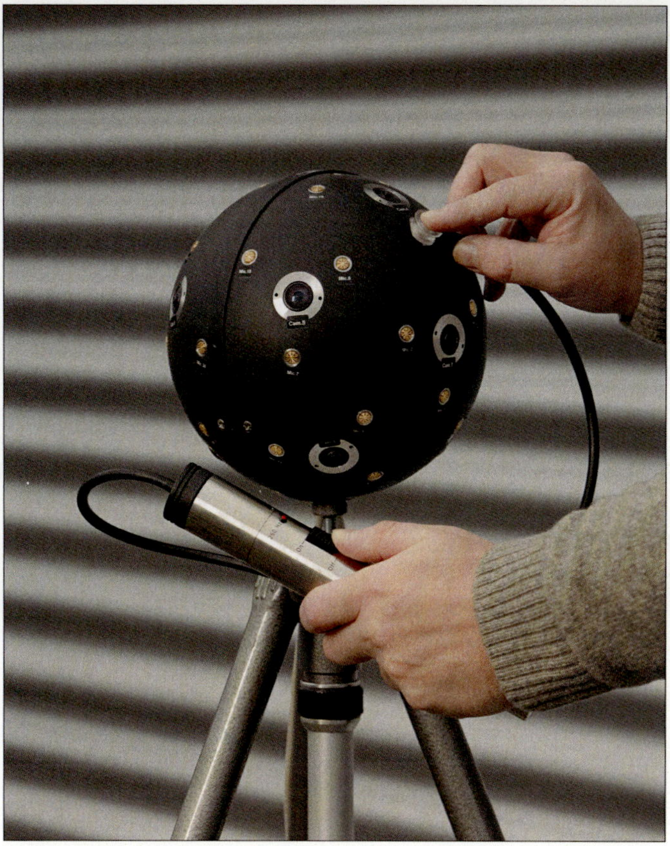

Fig. 1.3 The Brüel and Kjær spherical microphone array. This rigid array is comprised of 36 or 50 microphones and 12 video cameras. Copyright ©Brüel & Kjær. Used with permission

1.3 Organization of the Book

The content of this book is structured as follows:

- In **Chap.** 2, the fundamentals of acoustics are reviewed. We introduce the spherical harmonics, which form a complete set of orthonormal functions. Their importance rests in the fact that any arbitrary function on a sphere can be expanded in terms of a these functions and a set of expansion coefficients.
- **Chapter** 3 examines issues relating to spatial signal acquisition and transformation. We present the short-time Fourier transform and spherical harmonic framework that allow us to efficiently process the signals captured by a spherical microphone array. Common spatial sampling schemes are presented, which determine the placement of microphones on the sphere such that spatial aliasing is

minimized. In addition, we discuss the advantages and disadvantages of two common array types: the open and rigid arrays with omnidirectional microphones.

- In order to comprehensively evaluate spherical array processing algorithms under many different acoustic conditions, it is indispensable to use simulated acoustic impulse responses (AIRs). The image method proposed by Allen and Berkley [2] is a well-established way of doing this for point-to-point AIRs with sensors in free space, however it does not account for the scattering introduced by a rigid sphere. In **Chap.** 4, we present a method for simulating the AIRs between a sound source and microphones positioned on a rigid spherical array. In addition, three examples are presented based on this method: an analysis of a diffuse reverberant sound field, a study of binaural cues in the presence of reverberation, and an illustration of the algorithm's use as a mouth simulator.
- **Chapter** 5 introduces methods for the estimation of two important acoustic parameters: the DOA of a sound source, and the signal-to-diffuse energy ratio at a particular position in a sound field. Later in the book, it will be seen that these quantities can be used for signal enhancement purposes.
- The process of combining signals acquired by a microphone array in order to isolate a signal of interest is known as beamforming or spatial filtering. **Chapter** 6 considers the simplest type of beamformer: the signal-independent (fixed) beamformer, whose weights only depend on the DOA of the source to be extracted, and do not otherwise depend on the desired signal.
- In **Chap.** 7, we derive signal-dependent beamformers, whose weights depend on the second-order statistics of the desired signal and/or of the noise to be suppressed. These beamformers adaptively seek to achieve optimal performance in terms of noise reduction and speech distortion.
- **Chapter** 8 takes a different approach to signal enhancement: a physically-motivated parametric representation of the sound field is introduced. It is shown that the sound field can be manipulated to achieve noise reduction or dereverberation by applying a time- and frequency-dependent gain to a reference signal. The gain is a simple function of the sound field parameters, which can be estimated using the methods presented in Chap. 5.
- The concept of informed array processing is introduced in **Chap.** 9. It involves incorporating relevant spatial information about the specific problem into the design of spatial filters, and into the estimation of the second-order statistics that is required to implement the beamformers in Chap. 7. Informed array processing techniques are developed for two signal enhancement problems: noise reduction and dereverberation.

The structure of the book, and the relationship between each of the topics it addresses, is illustrated in Fig. 1.4.

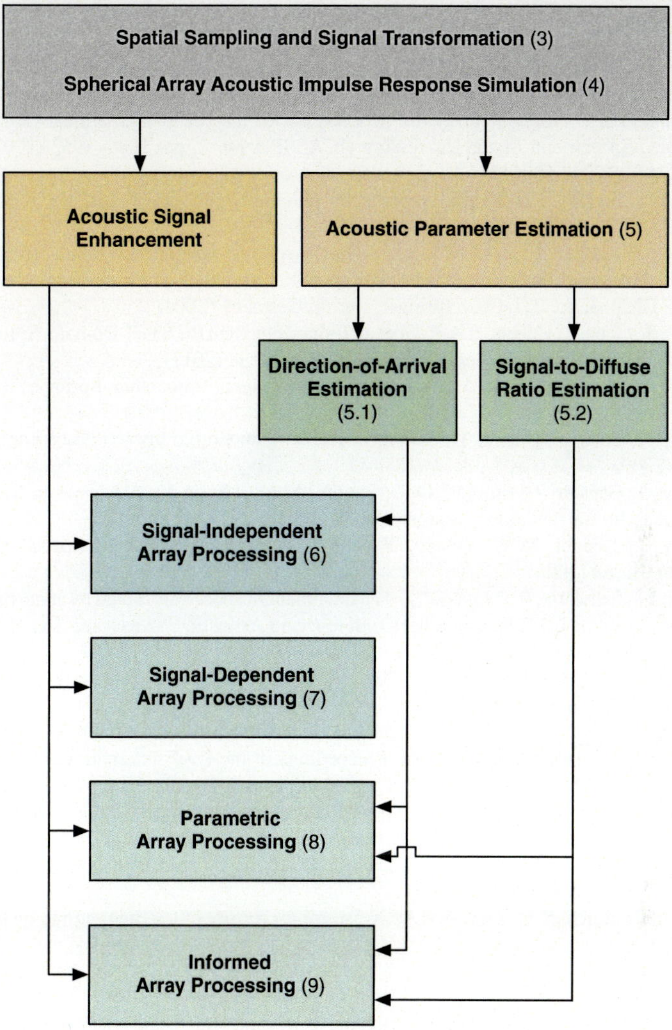

Fig. 1.4 Structure of the book. The chapter/section relating to each topic is indicated in *parentheses*

References

1. Abhayapala, T.D., Ward, D.B.: Theory and design of high order sound field microphones using spherical microphone array. In: Proceedings of the IEEE International Conference on Acoustics, Speech and Signal Processing (ICASSP), vol. 2, pp. 1949–1952 (2002). doi:10.1109/ICASSP.2002.1006151
2. Allen, J.B., Berkley, D.A.: Image method for efficiently simulating small-room acoustics. J. Acoust. Soc. Am. **65**(4), 943–950 (1979)
3. Assmann, P., Summerfield, Q.: The perception of speech under adverse conditions. In: Greenberg, S., Ainsworth, W.A., Popper, A.N., Fay, R.R. (eds.) Speech Processing in the Auditory System, Chap. 5, pp. 231–308. Springer, Berlin, Germany (2004)
4. Benesty, J., Chen, J., Habets, E.A.P.: Speech Enhancement in the STFT Domain. SpringerBriefs in Electrical and Computer Engineering. Springer, Berlin (2011)
5. Benesty, J., Chen, J., Huang, Y.: Microphone Array Signal Processing. Springer, Berlin, Germany (2008)
6. Benesty, J., Chen, J., Huang, Y., Cohen, I.: Noise Reduction in Speech Processing. Springer, Berlin (2009)
7. Benesty, J., Gänsler, T., Morgan, D.R., Sondhi, M.M., Gay, S.L.: Advances in Network and Acoustic Echo Cancellation. Springer, Berlin (2001)
8. Benesty, J., Sondhi, M.M., Huang, Y. (eds.): Springer Handbook of Speech Processing. Springer, Berlin (2008)
9. Berouti, M., Schwartz, R., Makhoul, J.: Enhancement of speech corrupted by acoustic noise. In: Proceedings of the IEEE International Conference on Acoustics, Speech and Signal Processing (ICASSP), vol. 4, pp. 208–211 (1979)
10. Brandstein, M.S., Ward, D.B. (eds.): Microphone Arrays: Signal Processing Techniques and Applications. Springer, Berlin (2001)
11. Braun, S., Jarrett, D.P., Fischer, J., Habets, E.A.P.: An informed spatial filter for dereverberation in the spherical harmonic domain. In: Proceedings of the IEEE International Conference on Acoustics, Speech and Signal Processing (ICASSP), pp. 669–673. Vancouver, Canada (2013)
12. Compton, Jr., R.: Adaptive Antennas, 1st edn. Prentice-Hall, Upper Saddle River (1988)
13. Doclo, S., Gannot, S., Moonen, M., Spriet, A.: Acoustic beamforming for hearing aid applications. In: Haykin, S., Liu, K.R. (eds.) Handbook on Array Processing and Sensor Networks, chap. 9. Wiley, New York (2008)
14. Eaton, J., Gaubitch, N.D., Naylor, P.A.: Noise-robust reverberation time estimation using spectral decay distributions with reduced computational cost. In: Proceedings of the IEEE International Conference on Acoustics, Speech and Signal Processing (ICASSP). Vancouver, Canada (2013)
15. Elko, G.W.: Future directions for microphone arrays. In: Brandstein and Ward [10], chap. 17, pp. 383–387
16. Elko, G.W., Meyer, J.: Spherical microphone arrays for 3D sound recordings. In: Huang, Y., Benesty, J. (eds.) Audio Signal Processing for Next-Generation Multimedia Communication Systems, chap. 3, pp. 67–89 (2004)
17. Elko, G.W., Meyer, J.: Microphone arrays. In: Benesty et al. [8], chap. 50
18. Gaubitch, N.D.: Blind identification of acoustic systems and enhancement of reverberant speech. Ph.D. thesis, Imperial College London (2006)
19. Gover, B.N., Ryan, J.G., Stinson, M.R.: Microphone array measurement system for analysis of directional and spatial variations of sound fields. J. Acoust. Soc. Am. **112**(5), 1980–1991 (2002). doi:10.1121/1.1508782
20. Gustafsson, T., Rao, B., Trivedi, M.: Source localization in reverberant environments: modeling and statistical analysis. IEEE Trans. Speech Audio Process. **11**(6), 791–803 (2003)
21. Habets, E.A.P.: Single- and multi-microphone speech dereverberation using spectral enhancement. Ph.D. thesis, Technische Universiteit Eindhoven (2007). http://alexandria.tue.nl/extra2/200710970.pdf

22. Habets, E.A.P., Benesty, J.: A perspective on frequency-domain beamformers in room acoustics. IEEE Trans. Audio, Speech, Lang. Process. **20**(3), 947–960 (2012)
23. Habets, E.A.P., Cohen, I., Gannot, S.: Generating nonstationary multisensor signals under a spatial coherence constraint. J. Acoust. Soc. Am. **124**(5), 2911–2917 (2008). doi:10.1121/1.2987429
24. Huang, Y., Benesty, J., Chen, J.: Dereverberation. In: Benesty et al. [8], chap. 5
25. Jarrett, D.P.: Spherical microphone array processing for acoustic parameter estimation and signal enhancement. Ph.D. thesis, Imperial College London (2013)
26. Jarrett, D.P., Habets, E.A.P., Benesty, J., Naylor, P.A.: A tradeoff beamformer for noise reduction in the spherical harmonic domain. In: Proceedings of the International Workshop on Acoust. Signal Enhancement (IWAENC). Aachen, Germany (2012)
27. Jarrett, D.P., Habets, E.A.P., Naylor, P.A.: 3D source localization in the spherical harmonic domain using a pseudointensity vector. In: Proceedings of the European Signal Processing Conference (EUSIPCO), pp. 442–446. Aalborg, Denmark (2010)
28. Jarrett, D.P., Habets, E.A.P., Naylor, P.A.: Spherical harmonic domain noise reduction using an MVDR beamformer and DOA-based second-order statistics estimation. In: Proceedings of the IEEE International Conference on Acoustics, Speech and Signal Processing (ICASSP), pp. 654–658. Vancouver, Canada (2013)
29. Jarrett, D.P., Habets, E.A.P., Thomas, M.R.P., Gaubitch, N.D., Naylor, P.A.: Dereverberation performance of rigid and open spherical microphone arrays: Theory & simulation. In: Proceedings of the Joint Workshop on Hands-Free Speech Communication and Microphone Arrays (HSCMA), pp. 145–150. Edinburgh, UK (2011)
30. Jarrett, D.P., Habets, E.A.P., Thomas, M.R.P., Naylor, P.A.: Simulating room impulse responses for spherical microphone arrays. In: Proceedings of the IEEE International Conference on Acoustics, Speech and Signal Processing (ICASSP), pp. 129–132. Prague, Czech Republic (2011)
31. Jarrett, D.P., Thiergart, O., Habets, E.A.P., Naylor, P.A.: Coherence-based diffuseness estimation in the spherical harmonic domain. In: Proceedings of the IEEE Convention of Electrical & Electronics Engineers in Israel (IEEEI). Eilat, Israel (2012)
32. Jeub, M., Nelke, C., Beaugeant, C., Vary, P.: Blind estimation of the coherent-to-diffuse energy ratio from noisy speech signals. In: Proceedings of the European Signal Processing Conf. (EUSIPCO). Barcelona, Spain (2011)
33. Kellermann, W.: Acoustic echo cancellation for beamforming microphone arrays. In: Brandstein, M.S., Ward, D.B. (eds.) Microphone Arrays: Signal Processing Techniques and Applications, pp. 281–306. Springer, Berlin, Germany (2001)
34. Khaykin, D., Rafaely, B.: Coherent signals direction-of-arrival estimation using a spherical microphone array: Frequency smoothing approach. In: Proceedings of the IEEE Workshop on Applications of Signal Processing to Audio and Acoustics, pp. 221–224 (2009). doi:10.1109/ASPAA.2009.5346492
35. Kuttruff, H.: Room Acoustics, 4th edn. Taylor & Francis, London (2000)
36. Li, Z., Duraiswami, R.: Flexible and optimal design of spherical microphone arrays for beamforming. IEEE Trans. Audio, Speech, Lang. Process. **15**(2), 702–714 (2007). doi:10.1109/TASL.2006.876764
37. Lim, F., Naylor, P.A.: Robust low-complexity multichannel equalization for dereverberation. In: Proceedings of the IEEE International Conference on Acoustics, Speech and Signal Processing (ICASSP). Vancouver, Canada (2013)
38. Lim, F., Thomas, M., Naylor, P.: Mintformer: A spatially aware channel equalizer. In: Proceedings of the IEEE Workshop on Applications of Signal Processing to Audio and Acoustics. New Paltz, USA (2013)
39. Löllmann, H., Vary, P.: Estimation of the frequency dependent reverberation time by means of warped filter-banks. In: Proceedings of the IEEE International Conference on Acoustics, Speech and Signal Processing (ICASSP), pp. 309 –312 (2011). doi:10.1109/ICASSP.2011.5946402

40. de M. Prego, T., de Lima, A.A., Netto, S.L., Lee, B., Said, A., Schafer, R.W., Kalker, T.: A blind algorithm for reverberation-time estimation using subband decomposition of speech signals. J. Acoust. Soc. Am. **131**(4), 2811–2816 (2012)
41. Meyer, J., Elko, G.: A highly scalable spherical microphone array based on an orthonormal decomposition of the soundfield. In: Proceedings of the IEEE International Conference on Acoustics, Speech and Signal Processing (ICASSP), vol. 2, pp. 1781–1784 (2002)
42. Naylor, P.A., Gaubitch, N.D. (eds.): Speech Dereverberation. Springer, Berlin (2010)
43. Pulkki, V.: Spatial sound reproduction with directional audio coding. J. Audio Eng. Soc. **55**(6), 503–516 (2007)
44. Rafaely, B.: Analysis and design of spherical microphone arrays. IEEE Trans. Speech Audio Process. **13**(1), 135–143 (2005). doi:10.1109/TSA.2004.839244
45. Rafaely, B., Peled, Y., Agmon, M., Khaykin, D., Fisher, E.: Spherical microphone array beamforming. In: I. Cohen, J. Benesty, S. Gannot (eds.) Speech Processing in Modern Communication: Challenges and Perspectives, chap. 11. Springer (2010)
46. Ratnam, R., Jones, D.L., Wheeler, B.C., O'Brien Jr., W.D., Lansing, C.R., Feng, A.S.: Blind estimation of reverberation time. J. Acoust. Soc. Am. **114**(5), 2877–2892 (2003)
47. Sondhi, M.: Adaptive echo cancelation for voice signals. In: Benesty et al. [8], chap. 45. Part H
48. Sun, H., Yan, S., Svensson, U.P.: Robust minimum sidelobe beamforming for spherical microphone arrays. IEEE Trans. Audio, Speech, Lang. Process. **19**(4), 1045–1051 (2011). doi:10.1109/TASL.2010.2076393
49. Talmon, R., Habets, E.A.P.: Blind reverberation time estimation by intrinsic modeling of reverberant speech. In: Proceedings of the IEEE International Conference on Acoustics, Speech and Signal Processing (ICASSP). Vancouver, Canada (2013)
50. Teutsch, H.: Wavefield decomposition using microphone arrays and its application to acoustic scene analysis. Ph.D. thesis, Friedrich-Alexander Universität Erlangen-Nürnberg (2005)
51. Teutsch, H., Kellermann, W.: EB-ESPRIT: 2D localization of multiple wideband acoustic sources using eigen-beams. In: Proceedings of the IEEE International Conference on Acoustics, Speech and Signal Processing (ICASSP), vol. 3, pp. iii/89–iii/92 (2005). doi:10.1109/ICASSP.2005.1415653
52. Teutsch, H., Kellermann, W.: Eigen-beam processing for direction-of-arrival estimation using spherical apertures. In: Proceedings of the Joint Workshop on Hands-Free Speech Communication and Microphone Arrays. Piscataway, New Jersey, USA (2005)
53. Teutsch, H., Kellermann, W.: Detection and localization of multiple wideband acoustic sources based on wavefield decomposition using spherical apertures. In: Proceedings of the IEEE International Conference on Acoustics, Speech and Signal Processing (ICASSP), pp. 5276–5279 (2008). doi:10.1109/ICASSP.2008.4518850
54. Thiergart, O., Del Galdo, G., Habets, E.A.P.: On the spatial coherence in mixed sound fields and its application to signal-to-diffuse ratio estimation. J. Acoust. Soc. Am. **132**(4), 2337–2346 (2012)
55. Thiergart, O., Del Galdo, G., Habets, E.A.P.: Signal-to-reverberant ratio estimation based on the complex spatial coherence between omnidirectional microphones. In: Proceedings of the IEEE International Conference on Acoustics, Speech and Signal Processing (ICASSP), pp. 309–312 (2012)
56. Wang, H., Kaveh, M.: Coherent signal-subspace processing for the detection and estimation of angles of arrival of multiple wide-band sources. IEEE Trans. Acoust., Speech, Signal Process. **33**(4), 823–831 (1985)
57. Wax, M.: Detection and localization of multiple sources via the stochastic signals model. IEEE Trans. Signal Process. **39**(11), 2450–2456 (1991)
58. Wen, J.Y.C., Habets, E.A.P., Naylor, P.A.: Blind estimation of reverberation time based on the distribution of signal decay rates. In: Proceedings of the IEEE International Conference on Acoustics, Speech and Signal Processing (ICASSP). Las Vegas, USA (2008)

Chapter 2
Theoretical Preliminaries of Acoustics

In this chapter, we review some of the fundamentals of acoustics and introduce the spherical harmonic expansion of a sound field, which is the basis for the spherical harmonic processing framework used with spherical microphone arrays.

This chapter intends to introduce the key theory and equations required in the rest of the book. For a more comprehensive introduction to acoustics, the reader is referred to [2, 12], or [17, 20] for a thorough treatment of acoustics in spherical coordinates.

2.1 Fundamentals of Acoustics

The propagation of acoustic waves through a material is described by a second-order partial differential equation known as the *wave equation*. The homogeneous wave equation describes the evolution of the sound pressure p as a function of time t and position $\llcorner\mathbf{r} = (x, y, z)$ in a homogeneous, source-free medium.[1] In three dimensions it is given by [12, Eq. 1.5]

$$\nabla^2 p(\llcorner\mathbf{r}, t) - \frac{1}{c^2}\frac{\partial^2 p(\llcorner\mathbf{r}, t)}{\partial t^2} = 0, \tag{2.1}$$

where

$$\nabla^2 = \frac{\partial^2}{\partial x^2} + \frac{\partial^2}{\partial y^2} + \frac{\partial^2}{\partial z^2} \tag{2.2}$$

[1]In this section, vectors in Cartesian coordinates are denoted with a corner mark $\llcorner$ to distinguish them from vectors in spherical coordinates, which will be introduced in Sect. 2.2.

D. Jarrett et al., *Theory and Applications of Spherical Microphone Array Processing*, Springer Topics in Signal Processing 9,
DOI 10.1007/978-3-319-42211-4_2

is the Laplace operator in Cartesian coordinates (x, y, z) and c denotes the speed of sound. The separation of variables method is used to simplify the analysis. The time-harmonic solution to the wave equation can then be written in the form

$$p({}_{\llcorner}\mathbf{r}, t) = P({}_{\llcorner}\mathbf{r}, k)e^{i\omega t}, \tag{2.3}$$

where $i = \sqrt{-1}$, and $P({}_{\llcorner}\mathbf{r}, k)$, to be defined later in this section, is a function of the position ${}_{\llcorner}\mathbf{r}$ and the wavenumber k. The wavenumber is related to the angular frequency ω, ordinary frequency f and speed of sound c via the dispersion relation

$$k = \frac{\omega}{c} = \frac{2\pi f}{c}. \tag{2.4}$$

The acoustic waves are assumed to be propagating in a non-dispersive medium, such that the propagation speed c is independent of the wavenumber k. Throughout this book, the speed of sound is assumed to be constant; when a numerical value is required, we will use $c = 343$ m/s, obtained when the medium is air at a temperature of approximately 19 °C [12, Eq. 1.1].

The function $P({}_{\llcorner}\mathbf{r}, k)e^{i\omega t}$ in (2.3) can be represented in the complex plane by a rotating vector or a *phasor*. The time-independent vector, represented by the complex number $P({}_{\llcorner}\mathbf{r}, k)$, is the *complex amplitude*. The complex amplitude is multiplied by the unit vector $e^{i\omega t}$ rotating anti-clockwise at speed ω (in rad $\cdot$ s^{-1}), which is the angular frequency of the harmonic function.

> **Warning:**
> Throughout this book, $e^{i\omega t}$ represents the time dependence of a positive-frequency wave; a convention that is commonly adopted in electrical and mechanical engineering. In Sect. 2.3, we will summarize the effect of the choice of convention on the key equations of this chapter.

The Fourier transform of a time-domain signal $f(t)$ is defined as

$$\mathcal{F}\{f(t)\} = \int_{-\infty}^{\infty} f(t)e^{-i\omega t}\mathrm{d}t. \tag{2.5}$$

As a consequence, the $e^{i\omega t}$ term in the time-harmonic solution to the wave equation (2.3) is eliminated when applying the Fourier transform. Using (2.5), the frequency-domain homogeneous wave equation, also known as the *homogeneous Helmholtz equation*, is obtained [12, Eq. 3.1]:

$$\nabla^2 P({}_{\llcorner}\mathbf{r}, k) + k^2 P({}_{\llcorner}\mathbf{r}, k) = 0, \tag{2.6}$$

where $P({}_{\llcorner}\mathbf{r}, k) = \mathcal{F}\{p({}_{\llcorner}\mathbf{r}, t)\}$ denotes the temporal Fourier transform of $p({}_{\llcorner}\mathbf{r}, t)$.

The homogeneous wave equation and Helmholtz equation assume a source-free medium. If waves are being produced by a harmonic disturbance, a source function of the form $s(\llcorner\mathbf{r}, t) = S(\llcorner\mathbf{r}, k)e^{i\omega t}$ is added to the right-hand side of the homogeneous wave equation (2.1) to obtain the *inhomogeneous* wave equation

$$\nabla^2 p(\llcorner\mathbf{r}, t) - \frac{1}{c^2}\frac{\partial^2 p(\llcorner\mathbf{r}, t)}{\partial t^2} = -s(\llcorner\mathbf{r}, t), \tag{2.7}$$

and by taking the temporal Fourier transform $\mathcal{F}$, we obtain the inhomogeneous Helmholtz equation

$$\nabla^2 P(\llcorner\mathbf{r}, k) + k^2 P(\llcorner\mathbf{r}, k) = -S(\llcorner\mathbf{r}, k). \tag{2.8}$$

In the presence of a unit-amplitude harmonic point source at a position $\llcorner\mathbf{r}_s$, the solution to the wave equation is known as the *Green's function* and is denoted by $G(\llcorner\mathbf{r}|\llcorner\mathbf{r}_s, k)$. Alternatively it is termed an *acoustic transfer function* (ATF) from the point $\llcorner\mathbf{r}_s$ to the point $\llcorner\mathbf{r}$. The frequency-domain source function is then given by $S(\llcorner\mathbf{r}, k) = \delta_3(\llcorner\mathbf{r} - \llcorner\mathbf{r}_s)$, where $\delta_3(\cdot)$ denotes the three dimensional Dirac delta function, and the Green's function can be found by solving the following equation:

$$\nabla^2 G(\llcorner\mathbf{r}|\llcorner\mathbf{r}_s, k) + k^2 G(\llcorner\mathbf{r}|\llcorner\mathbf{r}_s, k) = -\delta_3(\llcorner\mathbf{r} - \llcorner\mathbf{r}_s). \tag{2.9}$$

The Green's function must also satisfy a boundary condition at infinity, the *Sommerfeld radiation condition*, which ensures that sources radiate energy instead of absorbing it. It is given by [20, Eq. 8.28]

$$\lim_{||\llcorner\mathbf{r} - \llcorner\mathbf{r}_s|| \to \infty} ||\llcorner\mathbf{r} - \llcorner\mathbf{r}_s|| \left(\frac{\partial G(\llcorner\mathbf{r}|\llcorner\mathbf{r}_s, k)}{\partial ||\llcorner\mathbf{r} - \llcorner\mathbf{r}_s||} - ikG(\llcorner\mathbf{r}|\llcorner\mathbf{r}_s, k) \right) = 0, \tag{2.10}$$

where $||\cdot||$ denotes the 2-norm (Euclidean norm).

For a source at a position $\llcorner\mathbf{r}_s$ and a receiver at a position $\llcorner\mathbf{r}$, a solution to the inhomogeneous Helmholtz equation satisfying the Sommerfeld radiation condition is given by the *free-space Green's function*, where free-space indicates that the only boundary condition that applies is the Sommerfeld radiation condition, that is, the waves are not propagating within an enclosure. The free-space Green's function is given by [20, Eq. 8.5]

$$G(\llcorner\mathbf{r}|\llcorner\mathbf{r}_s, k) = \frac{e^{-ik||\llcorner\mathbf{r} - \llcorner\mathbf{r}_s||}}{4\pi||\llcorner\mathbf{r} - \llcorner\mathbf{r}_s||}. \tag{2.11}$$

From (2.11) it is clear that $G(\llcorner\mathbf{r}|\llcorner\mathbf{r}_s, k) = G(\llcorner\mathbf{r}_s|\llcorner\mathbf{r}, k)$. This equality represents one of the most fundamental examples of the principle of acoustic reciprocity because the pressure at a receiver point is unchanged when exchanging the source and receiver positions.

2.2 Sound Field Representation Using Spherical Harmonic Expansion

To describe the sound field on the surface of a sphere, we need to find the solutions of the Helmholtz differential equation, as described in the previous section, on the surface of the sphere. In the following, we introduce spherical harmonics, which are a series of special functions defined on the surface of a sphere and are commonly used to solve such differential equations. After introducing the spherical harmonics, we introduce a spherical harmonic expansion of the free-space Green's function that underpins the spherical harmonic domain (SHD) processing in this book.

We adopt the spherical coordinate system used in [5, 13, 18, 20], which is illustrated in Fig. 2.1. The spherical coordinates are related to Cartesian coordinates x, y, z via the expressions [20, Eq. 2.47]

$$x = r \sin\theta \cos\phi, \tag{2.12a}$$

$$y = r \sin\theta \sin\phi, \tag{2.12b}$$

$$z = r \cos\theta, \tag{2.12c}$$

where r, θ and ϕ respectively denote the radius, inclination and azimuth. Conversely, the spherical coordinates may be obtained from the Cartesian coordinates using

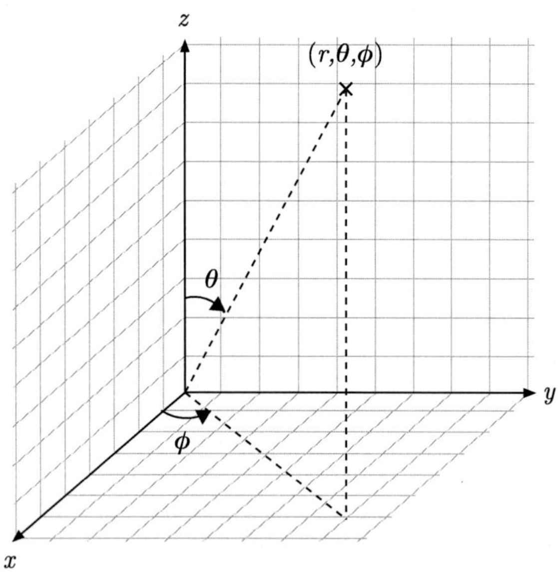

Fig. 2.1 Spherical coordinate system used in this book, defined relative to Cartesian coordinates. The radial distance r is the distance between the observation point and the origin of the coordinate system. The inclination angle θ (a.k.a. co-latitude, polar angle, or normal angle) is measured from the positive z-axis, and the azimuth angle ϕ is measured in the xy-plane from the positive x-axis.

$$r = \sqrt{x^2 + y^2 + z^2}, \tag{2.13a}$$

$$\theta = \arccos\left(\frac{z}{r}\right), \tag{2.13b}$$

$$z = \arctan\left(\frac{y}{x}\right), \tag{2.13c}$$

where arctan is the four-quadrant inverse tangent (implemented using the function `atan2()` in many computational environments including, for example, MATLAB).

We express the vectors $_{\mathrm{L}}\mathbf{r}$ and $_{\mathrm{L}}\mathbf{r}_{\mathrm{s}}$ in spherical coordinates as $\mathbf{r} = (r, \Omega) = (r, \theta, \phi)$ and $\mathbf{r}_{\mathrm{s}} = (r, \Omega_{\mathrm{s}})$. It is hereafter assumed that when the addition, scalar product and 2-norm operators are applied to vectors in spherical coordinates, these operations will in fact be performed in the Cartesian space by first performing a conversion from spherical to Cartesian coordinates using (2.12).

The *spherical harmonic* of order $l \geq 0$ and degree or mode m (satisfying $|m| < l$) is denoted by Y_{lm} and defined as

$$Y_{lm}(\Omega) = Y_{lm}(\theta, \phi) = \sqrt{\frac{(2l+1)}{4\pi}\frac{(l-m)!}{(l+m)!}}\mathcal{P}_{lm}(\cos\theta)e^{im\phi}, \tag{2.14}$$

where $\mathcal{P}_{lm}$ denotes the associated Legendre function of order[2] l and degree m.

The spherical harmonics, derived in [1, 20], represent the *angular component* of the solutions to the Helmholtz equation in spherical coordinates, and are involved in solving many problems in spherical coordinates. A number of zero-, first- and second-order spherical harmonics are plotted for illustrative purposes in Fig. 2.2.

For positive degrees m, the associated Legendre functions are related to the Legendre polynomials $\mathcal{P}_l(x)$ by the formula

$$\mathcal{P}_{lm}(x) = (-1)^m (1 - x^2)^{m/2} \frac{\mathrm{d}^m}{\mathrm{d}x^m}\mathcal{P}_l(x), \tag{2.15}$$

where the factor $(-1)^m$ is known as the Condon-Shortley phase. For negative degrees m, the associated Legendre functions can be obtained from

$$\mathcal{P}_{l(-m)}(x) = (-1)^m \frac{(l-m)!}{(l+m)!}\mathcal{P}_{lm}(x), \tag{2.16}$$

[2] In this book, for consistency with spherical array processing literature, we refer to l as the order and m as the degree of the spherical harmonics and associated Legendre functions (or polynomials). However, it should be noted that in other fields, l is referred to as the degree, and m as the order. This reflects the fact that the words *degree* and *order* are used interchangeably when referring to polynomials.

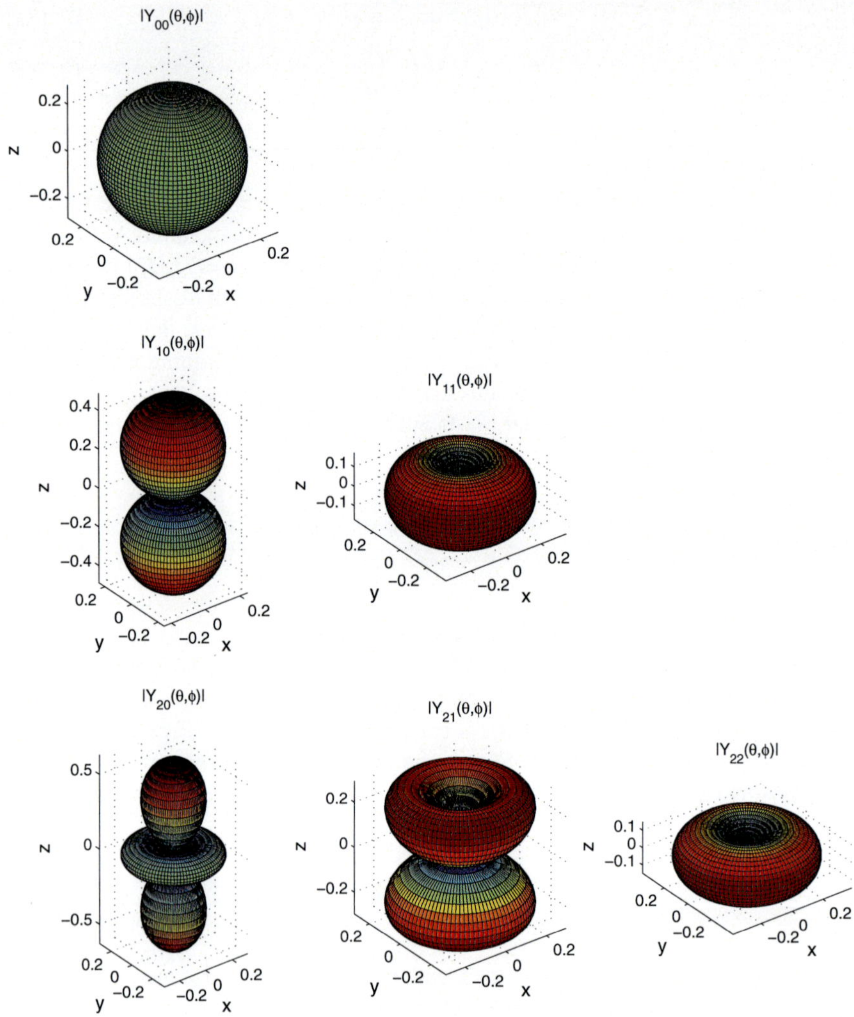

Fig. 2.2 Magnitude $|Y_{lm}(\theta, \phi)|$ of the complex spherical harmonics for $\{l \in \mathbb{Z} | 0 \leq l \leq 2\}$, $\{m \in \mathbb{Z} | 0 \leq m \leq l\}$. The plots for $m < 0$ are omitted as they are identical to those for $m > 0$. Copyright ©Daniel Jarrett. Used with permission

where $m > 0$. From (2.16) it follows that the spherical harmonics for corresponding negative degrees m can be computed using

$$Y_{l(-m)}(\Omega) = (-1)^m \, Y^*_{lm}(\Omega), \tag{2.17}$$

where $m > 0$.

The spherical harmonics constitute an *orthonormal* set of solutions to the Helmholtz equation in spherical coordinates, that is [20, Eq. 6.45]:

$$\int_{\Omega \in \mathcal{S}^2} Y_{lm}(\Omega) Y^*_{l'm'}(\Omega) \mathrm{d}\Omega = \delta_{l,l'} \delta_{m,m'}, \tag{2.18}$$

where the notation $\int_{\Omega \in \mathcal{S}^2} \mathrm{d}\Omega$ is used to denote compactly the solid angle[3] $\int_{\phi=0}^{2\pi} \int_{\theta=0}^{\pi} \sin\theta \mathrm{d}\theta \mathrm{d}\phi$, and the Kronecker delta $\delta_{i,j}$ is defined as

$$\delta_{i,j} = \begin{cases} 1, \text{ if } i = j; \\ 0, \text{ if } i \neq j. \end{cases} \tag{2.19}$$

In addition, they constitute a *complete* set of solutions, or equivalently they satisfy the completeness relation [20, Eq. 6.47]

$$\sum_{l=0}^{\infty} \sum_{m=-l}^{l} Y_{lm}(\theta, \phi) Y^*_{lm}(\theta', \phi') = \delta(\cos\theta - \cos\theta')\delta(\phi - \phi'), \tag{2.20}$$

where δ denotes the Dirac delta function. As a result, any function on a sphere can be represented using a *spherical harmonic expansion* (SHE).

In particular, the free-space Green's function (2.11) can be expanded using the following SHE [20, Eqs. 8.22 and 8.76]:

$$G(\mathbf{r}|\mathbf{r}_\mathrm{s}, k) = \frac{e^{-ik||\mathbf{r}-\mathbf{r}_\mathrm{s}||}}{4\pi||\mathbf{r} - \mathbf{r}_\mathrm{s}||} \tag{2.21}$$

$$= \sum_{l=0}^{\infty} \sum_{m=-l}^{l} \underbrace{-i\,k\,j_l(kr)\,h_l^{(2)}(kr_\mathrm{s}) Y^*_{lm}(\Omega_\mathrm{s})}_{\text{expansion coefficients}}\; Y_{lm}(\Omega), \tag{2.22}$$

where $(\cdot)^*$ denotes the complex conjugate, j_l is the spherical Bessel function of order l, and $h_l^{(2)}$ is the spherical Hankel function of the second kind and of order l. The spherical Bessel function forms the real part of the Hankel function, and the spherical Neumann function forms its imaginary part. The spherical Hankel function of the first kind $h_l^{(1)}$, used in Sect. 2.3, is the complex conjugate of $h_l^{(2)}$. The spherical Bessel and Neumann functions represent the *radial component* of the solutions to the Helmholtz equation in spherical coordinates.

In many cases it is convenient to remove the sum over all degrees m in (2.22) using the *spherical harmonic addition theorem* [1], which states that

$$\sum_{m=-l}^{l} Y^*_{lm}(\Omega_\mathrm{s}) Y_{lm}(\Omega) = \frac{2l+1}{4\pi} \mathcal{P}_l\left(\frac{\mathbf{r} \cdot \mathbf{r}_\mathrm{s}}{r r_\mathrm{s}}\right) \tag{2.23a}$$

$$= \frac{2l+1}{4\pi} \mathcal{P}_l(\cos\Theta), \tag{2.23b}$$

[3] The factor $\sin\theta$ compensates for the denser sampling near the poles ($\theta = 0$ and $\theta = \pi$).

where $\mathbf{r} \cdot \mathbf{r}_s$ denotes the scalar product[4] of the vectors $\mathbf{r}$ and $\mathbf{r}_s$, $\mathcal{P}_l$ is the Legendre polynomial of order l and Θ is the angle between $\mathbf{r}$ and $\mathbf{r}_s$. Using (2.23), the SHE of the free-space Green's function (2.22) then becomes

$$G(\mathbf{r}|\mathbf{r}_s, k) = -\frac{ik}{4\pi} \sum_{l=0}^{\infty} j_l(kr) h_l^{(2)}(kr_s)(2l+1)\mathcal{P}_l(\cos\Theta). \tag{2.24}$$

Under farfield conditions, when $r_s \to \infty$, the spherical wave represented by the free-space Green's functions (2.21) and (2.24) can be approximated as a plane wave. Although plane waves are only an approximation of spherical waves in the far field, plane waves are actually of utmost importance, since any complex wavefield can be represented as a superposition of plane waves [12, Chap. 1].

To obtain a far-field approximation of the Green's function given by (2.21), the denominator $||\mathbf{r} - \mathbf{r}_s||$ is first approximated by r_s. The phase term cannot be approximated so simply since it oscillates with respect to $||\mathbf{r} - \mathbf{r}_s||$. Instead, this term is approximated as [18, Eq. 2.29]

$$||\mathbf{r} - \mathbf{r}_s|| \approx r_s - \frac{\mathbf{r} \cdot \mathbf{r}_s}{r_s} \tag{2.25a}$$

$$\approx r_s - r\cos\Theta. \tag{2.25b}$$

Applying these approximations to (2.21) then yields

$$G(\mathbf{r}|\mathbf{r}_s, k) = \frac{e^{-ik||\mathbf{r}-\mathbf{r}_s||}}{4\pi||\mathbf{r} - \mathbf{r}_s||} \approx \frac{e^{-ikr_s}}{4\pi r_s} e^{+ikr\cos\Theta}. \tag{2.26}$$

A farfield approximation for the SHE of the Green's function given by (2.24) can be obtained by making use of the large argument approximation of the spherical Hankel function [20, Eqs. 6.68 and 6.58]:

$$h_l^{(2)}(kr_s) \approx i^{l+1} \frac{e^{-ikr_s}}{kr_s} \quad \text{for} \quad kr_s \gg 1 \tag{2.27}$$

Applying this approximation to the free-space Green's function (2.24), we obtain

$$G(\mathbf{r}|\mathbf{r}_s, k) \approx \frac{e^{-ikr_s}}{4\pi r_s} \sum_{l=0}^{\infty} i^l j_l(kr)(2l+1)\mathcal{P}_l(\cos\Theta). \tag{2.28}$$

[4]As noted earlier in the chapter, the scalar product of vectors in spherical coordinates is applied after these vectors have been converted to Cartesian coordinates using (2.12).

The second term in (2.28) is equal to $e^{+ikr\cos\Theta}$ [20, Eq. 6.174], that is,

$$e^{+ikr\cos\Theta} = \sum_{l=0}^{\infty} i^l j_l(kr)(2l+1)\mathcal{P}_l(\cos\Theta). \tag{2.29}$$

This is the expression for the pressure measured at a position **r** due to a unit-amplitude plane wave incident from a direction Ω_s, which we will use on multiple occasions later in this book.

2.3 Sign Convention

As mentioned in Sect. 2.1 the way the time dependence of a positive-frequency wave is defined affects many of the equations in this chapter, such as (2.3). In order to avoid any confusion, we have listed the key equations under each convention in Table 2.1.

Table 2.1 Key equations under the two sign conventions: one common in the acoustics literature and the other common in the engineering literature. The engineering convention is adopted in this book

Acoustics convention	Engineering convention
Temporal Fourier transform	
$\mathcal{F}\{f(t)\} = \int_{-\infty}^{\infty} f(t)e^{+i\omega t}\mathrm{d}t$	$\mathcal{F}\{f(t)\} = \int_{-\infty}^{\infty} f(t)e^{-i\omega t}\mathrm{d}t$
Free-space Green's function	
$G(\mathbf{r}\vert\mathbf{r}_s, k) = \frac{e^{+ik\vert\vert\mathbf{r}-\mathbf{r}_s\vert\vert}}{4\pi\vert\vert\mathbf{r}-\mathbf{r}_s\vert\vert}$	$G(\mathbf{r}\vert\mathbf{r}_s, k) = \frac{e^{-ik\vert\vert\mathbf{r}-\mathbf{r}_s\vert\vert}}{4\pi\vert\vert\mathbf{r}-\mathbf{r}_s\vert\vert}$
Free-space Green's function (expansion)	
$G(\mathbf{r}\vert\mathbf{r}_s, k) = \frac{ik}{4\pi}\sum_{l=0}^{\infty} j_l(kr)h_l^{(1)}(kr_s) \times(2l+1)\mathcal{P}_l(\cos\Theta)$	$G(\mathbf{r}\vert\mathbf{r}_s, k) = \frac{-ik}{4\pi}\sum_{l=0}^{\infty} j_l(kr)h_l^{(2)}(kr_s) \times(2l+1)\mathcal{P}_l(\cos\Theta)$
Farfield approximation for the free-space Green's function	
$G(\mathbf{r}\vert\mathbf{r}_s, k) \approx \frac{e^{+ikr_s}}{4\pi r_s}e^{-ikr\cos\Theta}$	$G(\mathbf{r}\vert\mathbf{r}_s, k) \approx \frac{e^{-ikr_s}}{4\pi r_s}e^{+ikr\cos\Theta}$
Farfield approximation for the free-space Green's function (expansion)	
$G(\mathbf{r}\vert\mathbf{r}_s, k) \approx \frac{e^{+ikr_s}}{4\pi r_s}\sum_{l=0}^{\infty}(-i)^l j_l(kr) \times(2l+1)\mathcal{P}_l(\cos\Theta)$	$G(\mathbf{r}\vert\mathbf{r}_s, k) \approx \frac{e^{-ikr_s}}{4\pi r_s}\sum_{l=0}^{\infty} i^l j_l(kr) \times(2l+1)\mathcal{P}_l(\cos\Theta)$
Plane wave	
$e^{-ikr\cos\Theta} = \sum_{l=0}^{\infty}(-i)^l j_l(kr)(2l+1) \times\mathcal{P}_l(\cos\Theta)$	$e^{+ikr\cos\Theta} = \sum_{l=0}^{\infty} i^l j_l(kr)(2l+1) \times\mathcal{P}_l(\cos\Theta)$

The engineering convention is adopted in this book. A more detailed discussion of sign conventions in spherical microphone array processing and, in particular, the effects of inconsistent use of sign conventions, can be found in [19].

2.4 Sound Intensity

In acoustic signal processing, the signals of interest are usually sound pressure signals, the sound pressure being the physical quantity that is perceived by the ear. However, a sound field can also be analyzed in terms of the acoustic energy that is radiated, transmitted and absorbed [4], allowing sound sources to be located and their power to be determined.

The *sound intensity vector* describes the magnitude and direction of the flow of acoustic energy per unit area, and has units of watts per square metre. The *instantaneous* sound intensity vector at a position $\mathbf{r}$ and time t is defined as [12, Eq. 1.26]

$$\mathcal{I}(\mathbf{r}, t) = p(\mathbf{r}, t)\mathbf{v}(\mathbf{r}, t), \tag{2.30}$$

where $p(\mathbf{r}, t)$ and $\mathbf{v}(\mathbf{r}, t)$ respectively denote the sound pressure and particle velocity vector at a position $\mathbf{r}$.

The *time-averaged* intensity vector has been found to be of more practical significance [4], and is defined as [4, 20]

$$\mathcal{I}(\mathbf{r}) = \langle p(\mathbf{r}, t)\mathbf{v}(\mathbf{r}, t) \rangle , \tag{2.31}$$

where $\langle \cdot \rangle$ denotes the time-averaging operation. The time-average of the net flow of energy out of a closed surface $\mathcal{S}$ is zero unless power is generated (or dissipated) within this surface [4], in which case it is equal to the power P_{src} of the sound source enclosed, or equivalently [4, Eq. 5]

$$\oint_{\mathcal{S}} \mathcal{I} \cdot \mathrm{d}\mathbf{S} = P_{\text{src}}, \tag{2.32}$$

where $\mathrm{d}\mathbf{S}$ denotes the differential surface area vector normal to $\mathcal{S}$.

For a simple harmonic sound field with constant angular frequency ω, the time-averaged sound intensity vector can be expressed in complex notation as [4, 20]

$$\mathcal{I}(\mathbf{r}, \omega) = \frac{1}{2}\Re \left\{ p(\mathbf{r}, \omega)\mathbf{v}^*(\mathbf{r}, \omega) \right\}, \tag{2.33}$$

where $p(\mathbf{r}, \omega)$ and $\mathbf{v}(\mathbf{r}, \omega)$ are complex exponential quantities, and $\Re\{\cdot\}$ denotes the real part of a complex number.

In general, there is no simple relationship between the intensity vector and sound pressure [7]. Nevertheless, for a plane progressive wave, the sound pressure p is related to the particle velocity $\mathbf{v}$ via the relationship [14–16]

$$\mathbf{v}(\mathbf{r}, t) = -\frac{p(\mathbf{r}, t)}{\rho_0\, c}\mathbf{u}(\mathbf{r}, t), \tag{2.34}$$

where ρ_0 and c respectively denote the ambient density of the medium and speed of sound, and $\mathbf{u}$ is a unit vector pointing from $\mathbf{r}$ towards the source. In this case, the direction of arrival of a sound source can be determined as the direction opposite to that of the intensity vector $\mathcal{I}$.

Point sources produce spherical waves, but when sufficiently far from these sources, in the farfield, these waves can be considered as plane waves so that p and $\mathbf{v}$ are in phase. In contrast, in the nearfield, p and $\mathbf{v}$ are out of phase [4]. This phase relationship can be described by next introducing the concept of *active* and *reactive* sound fields. All time-stationary fields can be split into two components, described by [6]:

- An active intensity vector, given by the product of the pressure p and the in-phase component of the particle velocity vector $\mathbf{v}$, which is the intensity vector we have described thus far. The active intensity vector has a non-zero time average [8], computed using (2.33).
- A reactive intensity vector, given by the product of the pressure p and the out-of-phase component of the particle velocity vector $\mathbf{v}$, which measures the energy stored in a sound field. The time-average of the reactive intensity vector is zero; to quote Fahy: there is "*local oscillatory transport of energy*" [6].

In the nearfield, the reactive field is stronger than the active field [4, 11]. In an anechoic environment, where there are no reflections, the strength of the reactive field decreases rapidly as the distance from the source increases [4, 11], such that in the farfield the sound field is essentially an active field. In Chap. 5, we will also take advantage of the fact that in a diffuse sound field, often used to model reverberation, the time-averaged active intensity vector is zero [9].

In practice, measurement of the intensity vector is difficult: typically it is measured with two closely-spaced matched pressure microphones using a finite-difference approximation of the pressure gradient (the *p*–*p* method [10]), although this method is very sensitive to mismatches in the phase response of the two microphones. The alternative (the *p*-*u* method [10]) is to combine a pressure transducer and a particle velocity transducer; this can be done using the Microflown [3]. In Sect. 5.1.3, we will see that the intensity vector can also be measured using a spherical microphone array.

2.5 Chapter Summary

The main aim of this chapter has been to introduce some of the relevant elements of the fundamentals of acoustics. The chapter reviewed the key equations that govern the propagation of sound waves in a medium, more specifically, the wave equation,

Helmholtz equation, and free-space Green's function. We also presented the SHE of the Green's function; the SHE forms the basis of a processing framework that advantageously exploits the spherical symmetry of spherical microphone arrays. Finally, we introduced the sound intensity vector, which describes the magnitude and direction of the flow of acoustic energy. In Chap. 5, it will be seen that the intensity vector can be employed in the estimation of two acoustic parameters: the direction of arrival (DOA) of a sound source, and the diffuseness of a sound field.

References

1. Arfken, G.B., Weber, H.J.: Mathematical Methods for Physicists, 5th edn. Academic Press, San Diego (2001)
2. Beranek, L.L.: Acoustics. McGraw-Hill, New York (1954)
3. de Bree, H.E., Leussink, P., Korthorst, T., Jansen, H., Lammerink, T.S., Elwenspoek, M.: The μ-flown: A Novel Device for Measuring Acoustic Flows, pp. 552–557. Elsevier, Amsterdam (1996)
4. Crocker, M.J., Jacobsen, F.: Sound Intensity, Chap. 156, pp. 1855–1868. Wiley-Interscience, New York (1997)
5. Elko, G.W., Meyer, J.: Spherical microphone arrays for 3D sound recordings. In: Y. Huang, J. Benesty (eds.) Audio Signal Processing for Next-Generation Multimedia Communication Systems, Chap. 3, pp. 67–89 (2004)
6. Fahy, F.J.: Sound Intensity, first edn. Academic Press, Cambridge (1989)
7. Fahy, F.J.: Foundations of Engineering Acoustics, first edn. Academic Press, Cambridge (2001)
8. Hansen, C., Snyder, S., Qiu, X., Brooks, L., Moreau, D.: Active Control of Noise and Vibration, second edn. CRC Press, Boca Raton (2013)
9. Jacobsen, F.: Active and reactive sound intensity in a reverberant sound field. J. Sound Vib. **143**(2), 231–240 (1990). doi:10.1016/0022-460X(90)90952-V
10. Jacobsen, F., de Bree, H.E.: Measurement of sound intensity: p–u probes versus p–p probes. In: Proceedings of Noise and Vibration Emerging Methods (2005)
11. Jacobsen, F., Juhl, P.M.: Fundamentals of General Linear Acoustics. Wiley, New York (2013)
12. Kuttruff, H.: Room Acoustics, 4th edn. Taylor and Francis, London (2000)
13. Meyer, J., Agnello, T.: Spherical microphone array for spatial sound recording. In: Proceedings Audio Engineering Society Convention, pp. 1–9. New York (2003)
14. Morse, P.M., Ingard, K.U.: Theoretical Acoustics. International Series in Pure and Applied Physics. McGraw Hill, New York (1968)
15. Nehorai, A., Paldi, E.: Acoustic vector-sensor array processing. IEEE Trans. Signal Process. **42**(9), 2481–2491 (1994). doi:10.1109/78.317869
16. Pierce, A.D.: Acoustics: An Introduction to Its Physical Principles and Applications. Acoustical Society of America (1991)
17. Rafaely, B.: Fundamentals of Spherical Array Processing. Springer, Berlin (2015)
18. Teutsch, H.: Wavefield decomposition using microphone arrays and its application to acoustic scene analysis. Ph.D. thesis, Friedrich-Alexander Universität Erlangen-Nürnberg (2005)
19. Tourbabin, V., Rafaely, B.: On the consistent use of space and time conventions in array processing. Acta Acustica united with Acustica **101**(3), 470–473 (2015)
20. Williams, E.G.: Fourier Acoustics: Sound Radiation and Nearfield Acoustical Holography, 1st edn. Academic Press, London (1999)

Chapter 3
Spatial Sampling and Signal Transformation

Many publications that propose algorithms for parameter estimation or signal enhancement purposes begin from the outset using signals in either or both of the time-frequency and spherical harmonic domains. One aim of the current chapter is to provide some of the algorithmic details necessary to process signals directly from the microphones, which will then enable subsequent spherical harmonic domain processing to be applied.

The steps required to acquire and process signals from spherical microphone arrays are discussed in this chapter. As a first step, the sound field must be sampled using a spherical array composed of microphones arranged in a particular configuration. The acquired signals can then be transformed to the time-frequency domain, and subsequently transformed to the spherical harmonic domain, where they can be conveniently manipulated. As both transforms are linear, the order of the transforms can be exchanged. In this chapter, we first review the short-time Fourier transform and subsequently two alternative ways to obtain signals in the time-frequency domain and in the spherical harmonic domain. Finally, a number of microphone positioning schemes and array configurations are discussed.

3.1 Time-Frequency Domain Processing

In acoustic signal processing, it is common to analyze the acquired microphone signals using a time-frequency representation. This is appropriate because the signals of interest usually have time-varying spectral characteristics. In addition, speech signals are to some degree sparse in the time-frequency domain, that is, the majority

---

Portions of Sect. 3.4 were first published in [8], and are reproduced here with the author's permission.

D. Jarrett et al., *Theory and Applications of Spherical Microphone Array Processing*, Springer Topics in Signal Processing 9,
DOI 10.1007/978-3-319-42211-4_3

of the speech energy is contained in a small number of time-frequency bins; this property is exploited in the presence of multiple speech sources, allowing us to assume that only a single speaker is active in a single time-frequency bin [24].

The most common type of time-frequency analysis technique is the *short-time Fourier transform* (STFT), which we will adopt for this book. With this technique, the signal to be analyzed is divided into short, overlapping frames, an analysis window is applied to each of these frames, and the Fourier transform is applied.

The (forward) STFT [1] of a discrete-time signal $p[n]$, with n denoting discrete time, is given by the spectral coefficients

$$P(\ell, \nu) = \sum_{n=0}^{K-1} p\,[n + \ell N_{\mathrm{f}}]\,\psi[n] e^{-i\frac{2\pi}{K}\nu n}, \tag{3.1}$$

where ℓ is the time index, $0 \le \nu \le K - 1$ is the frequency index, $\psi[n]$ is the analysis window of length K, and N_{f} is the number of samples between successive frames and termed the frame step. The continuous frequency f is related to the frequency index ν via the expression $f = \frac{\nu}{K} f_{\mathrm{s}}$, for $0 \le \nu \le K/2$, where f_{s} is the sampling frequency. Using the dispersion relation (2.4), the wavenumber k can be expressed as

$$k = \frac{2\pi f}{c} = \frac{2\pi\nu f_{\mathrm{s}}}{cK}, \tag{3.2}$$

where c is the speed of sound. In practice, since the pressure signals to be analyzed are real, their spectral coefficients are conjugate symmetric. The spectral coefficients for $\nu = K/2 + 1, \ldots, K - 1$ are therefore obtained as $P(\ell, \nu) = P(\ell, K - \nu)^*$ using the coefficients for $\nu = 1, \ldots, K/2 - 1$.

The choice of analysis window, and in particular its length, determines the time and frequency resolution of the STFT: a short window provides high resolution in time, but low resolution in frequency, whereas a long window provides high resolution in frequency at the expense of low resolution in time.

In acoustic parameter estimation, only the forward STFT is typically used. In acoustic signal enhancement, we wish to reconstruct the original time-domain signal with an inverse STFT. To achieve this objective, normally following speech enhancement processing, the time-domain signal $p[n]$ can be synthesized from its spectral coefficients $P(\ell, \nu)$ as [25]

$$p[n] = \sum_{\ell} \sum_{\nu=0}^{K-1} P(\ell, \nu)\tilde{\psi}\,[n - \ell N_{\mathrm{f}}]\, e^{+i\frac{2\pi}{K}\nu(n-\ell N_{\mathrm{f}})} \tag{3.3}$$

where $\tilde{\psi}[n]$ denotes the synthesis window. The equality in (3.3) is actually an approximation in most cases. In order to achieve *perfect reconstruction*, the analysis and synthesis windows must satisfy the completeness condition [25]

$$\sum_{\ell} \psi[n - \ell N_{\mathrm{f}}]\tilde{\psi}[n - \ell N_{\mathrm{f}}] = 1, \forall n \in \mathbb{Z}. \tag{3.4}$$

The reconstruction is only error-free, however, if the spectral coefficients are not modified [25]. In practice, when performing acoustic signal enhancement, we wish to filter the acquired signals by applying a gain to their spectral coefficients. In this case, the objective is *near perfect reconstruction*. When no synthesis window is applied, or equivalently, when a rectangular window $\tilde{\psi}[n] = 1, \forall n \in \mathbb{Z}$ is applied, the method described in (3.3) is referred to as *overlap-add*. However, the rectangular window has high sidelobe levels, which cause large amounts of spectral leakage or *aliasing*. The use of a synthesis window minimizes this effect; the method is then referred to as *weighted overlap-add* (WOLA) [25].

The specific choice of the analysis and synthesis windows, along with the frame step N_{f}, is beyond the scope of this book. The reader is referred to [25] for an extensive discussion of these choices. A MATLAB implementation of WOLA is available in the VOICEBOX speech processing toolbox [5], in the form of two functions `enframe()` and `overlapadd()`.

3.2 Complex Spherical Harmonic Domain Processing

In Sect. 2.2, we showed that the spherical harmonics $Y_{lm}(\Omega)$ can be used to represent the solutions to the wave equation. In fact, as the spherical harmonics define a complete basis set on the sphere, any square-integrable function on a sphere $\chi(\Omega) = \chi(\theta, \phi)$ can be represented using a spherical harmonic expansion (SHE) [7]:

$$\chi(\theta, \phi) = \sum_{l=0}^{\infty} \sum_{m=-l}^{l} \chi_{lm} Y_{lm}(\theta, \phi), \tag{3.5}$$

where l and m respectively denote the order and degree (or mode) of the spherical harmonic expansion coefficients χ_{lm}. As the spherical harmonics are orthonormal, as seen in (2.18), these coefficients can be computed as

$$\chi_{lm} = \int_{\Omega \in \mathcal{S}^2} \chi(\theta, \phi) Y_{lm}^{*}(\Omega) \mathrm{d}\Omega, \tag{3.6}$$

where the notation $\int_{\Omega \in \mathcal{S}^2} \mathrm{d}\Omega$ is used to denote compactly the solid angle $\int_{\phi=0}^{2\pi} \int_{\theta=0}^{\pi} \sin\theta \mathrm{d}\theta \mathrm{d}\phi$.

The expansion in (3.5) is the spherical equivalent of the one-dimensional Fourier series. It is referred to as an inverse complex *spherical harmonic transform* (SHT) operation, while (3.6) is referred to as a forward complex SHT. This transform converts spatial domain signals to the spherical harmonic domain (SHD).

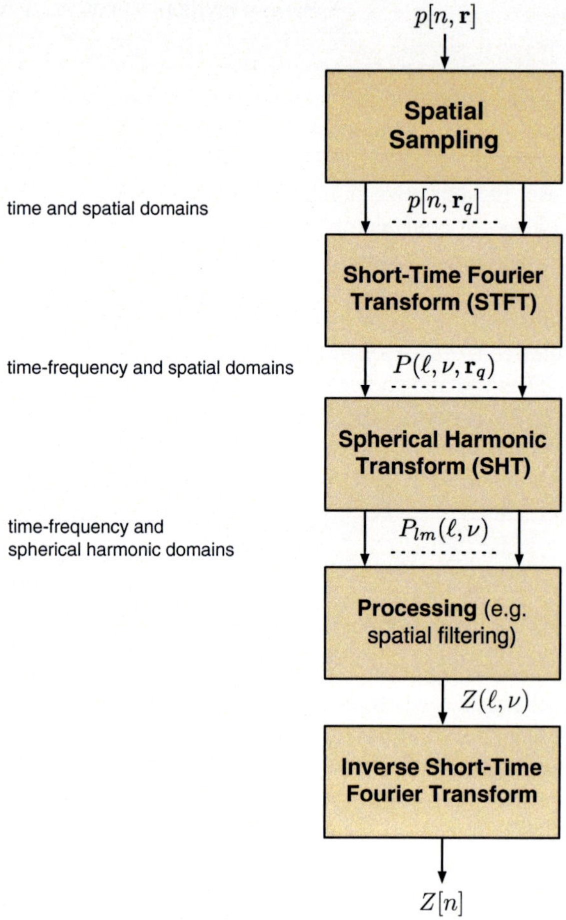

Fig. 3.1 Processing framework for spherical microphone arrays. The sound field is first sampled using a discrete set of Q microphones at positions $\mathbf{r}_q, q \in \{1, \ldots, Q\}$, and then transformed to the time-frequency domain and to the spherical harmonic domain

In this book, where we are specifically interested in acoustic signals, we choose to apply first the STFT to the signals captured using our spherical microphone array, and then apply the SHT, as illustrated in Fig. 3.1. The complex SHT of an STFT domain signal $P(\ell, \nu, \mathbf{r})$ captured at a position $\mathbf{r} = (r, \Omega)$ is given by

$$P_{lm}(\ell, \nu) = \int_{\Omega \in \mathcal{S}^2} P(\ell, \nu, \mathbf{r}) Y_{lm}^*(\Omega) \mathrm{d}\Omega, \tag{3.7}$$

while the inverse complex SHT is given by

$$P(\ell, \nu, \mathbf{r}) = \sum_{l=0}^{\infty} \sum_{m=-l}^{l} P_{lm}(\ell, \nu) Y_{lm}(\Omega). \tag{3.8}$$

With an acoustic signal, the SHD coefficients P_{lm} are often called *eigenbeams* to reflect the fact that the spherical harmonics are eigensolutions of the wave equation in spherical coordinates [26], or the fact that the eigenbeams characterize the sound field in a similar way as eigenvectors characterize a matrix [13].

It is interesting to note that the eigenbeams used in SHD microphone array processing can be viewed as single microphone signals in a classical microphone array signal processing sense [26]. While the relationship between the eigenbeams is different from the relationship between spatial domain microphone signals, and each eigenbeam is actually computed based on *all* the microphones in the array, many classical array processing algorithms can be adapted to the SHD, as we will see in Chaps. 5 and 7, for example.

3.3 Real Spherical Harmonic Domain Processing

In Sect. 3.2, signals that are in the time-frequency domain and in the SHD were obtained by first applying the STFT and then applying the SHT. However, the SHT can alternatively be applied before the STFT. In this case, the processing framework of Fig. 3.1 is replaced with the alternative processing framework of Fig. 3.2. This alternative framework can be advantageous in terms of computational complexity, as we will see in Sect. 3.4.

As the discrete time domain acoustic signals $p[n]$ are real, we apply the *real* SHT instead of the complex SHT. The real SHT of a discrete time domain signal $p[n, \mathbf{r}]$ captured at a position $\mathbf{r}$ is given by

$$p_{lm}[n] = \int_{\Omega \in \mathcal{S}^2} p[n, \mathbf{r}] R_{lm}(\Omega) \mathrm{d}\Omega, \tag{3.9}$$

while the inverse real SHT is given by

$$p[n, \mathbf{r}] = \sum_{l=0}^{\infty} \sum_{m=-l}^{l} p_{lm}[n] R_{lm}(\Omega), \tag{3.10}$$

where $R_{lm}(\Omega)$ denotes the *real* spherical harmonic of order l and degree m evaluated at an angle Ω.

The real-valued spherical harmonics can be defined by mapping the sine portion of the complex exponential $e^{im\phi} = \cos(m\,\phi) + i\,\sin(m\,\phi)$ to negative degrees m, the cosine portion to positive degrees m, and DC to $m = 0$, i.e.,

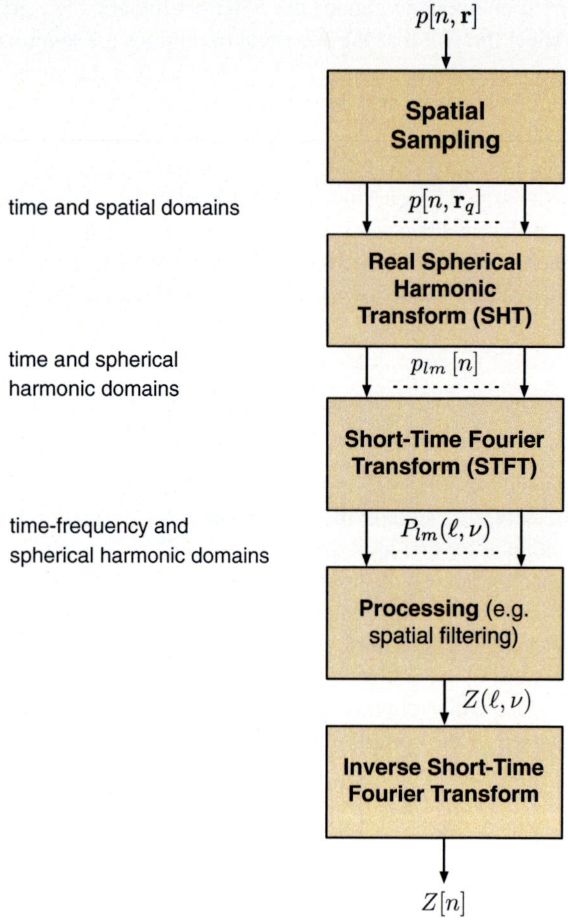

Fig. 3.2 Alternative processing framework for spherical microphone arrays. The sound field is first sampled, and then transformed to the SHD and to the time-frequency domain

$$R_{lm}(\Omega) = (-1)^{|m|} \sqrt{\frac{2l+1}{4\pi} \frac{(l-|m|)!}{(l+|m|)!}} \times \mathcal{P}_{l|m|}(\cos\theta) \begin{cases} \sqrt{2}\sin(|m|\,\phi) & m < 0 \\ 1 & m = 0 \\ \sqrt{2}\cos(m\,\phi) & m > 0 \end{cases} \tag{3.11}$$

where $|\cdot|$ denotes the absolute value and $\mathcal{P}_{lm}$ denotes the associated Legendre function of order l and degree m as defined in Sect. 2.2. Here the Condon–Shortley phase factor $(-1)^m$ is required to cancel the factor that is present in the definition of the associated Legendre functions. The real spherical harmonics as defined in (3.11) are most commonly used in the context of Ambisonics, and have the same orthonormality property (2.18) as the complex spherical harmonics defined in Chap. 2. It should be noted that, in principle, the cosine and sine portions can alternatively be mapped the other way around. The advantage of this alternative mapping is that the cosine is an even function and therefore the sign of m does not matter [9].

The real spherical harmonics $R_{lm}(\Omega)$ can equivalently be deduced from the complex spherical harmonics $Y_{lm}(\Omega)$ by extracting their real and imaginary parts and renormalizing as follows:

$$R_{lm}(\Omega) = \begin{cases} \sqrt{2}\,(-1)^m\, \Im\{Y_{l(-m)}(\Omega)\} & m < 0 \\ Y_{l0}(\Omega) & m = 0 \\ \sqrt{2}\,(-1)^m\, \Re\{Y_{lm}(\Omega)\} & m > 0 \end{cases} \tag{3.12}$$

where $\Re\{\cdot\}$ and $\Im\{\cdot\}$ respectively denote the real and imaginary parts of a complex number. The real spherical harmonics can also be expressed in terms of the complex spherical harmonics as follows

$$R_{lm}(\Omega) = \begin{cases} \frac{i}{\sqrt{2}}\left(Y_{lm}(\Omega) - (-1)^m\, Y_{l(-m)}(\Omega)\right) & m < 0 \\ Y_{l0}(\Omega) & m = 0\;. \\ \frac{1}{\sqrt{2}}\left(Y_{l(-m)}(\Omega) + (-1)^m\, Y_{lm}(\Omega)\right) & m > 0 \end{cases} \tag{3.13}$$

Finally, the complex spherical harmonics can be expressed in terms of the real spherical harmonics as follows

$$Y_{lm}(\Omega) = \begin{cases} \frac{1}{\sqrt{2}}\left(R_{l(-m)}(\Omega) - i\, R_{lm}(\Omega)\right) & m < 0 \\ Y_{l0}(\Omega) & m = 0\;. \\ \frac{1}{\sqrt{2}}(-1)^m\left(R_{lm}(\Omega) + i\, R_{l(-m)}(\Omega)\right) & m > 0 \end{cases} \tag{3.14}$$

Once the real SHT coefficients $p_{lm}[n]$ have been computed, the STFT can be applied to obtain time-frequency and spherical harmonic domain signals $P_{lm}(\ell, \nu)$, as explained in Sect. 3.1.

In the rest of this book, and in particular in Chaps. 5 to 9, the complex SHT will be used. If instead the real SHT is used, it is important to replace the complex spherical harmonics Y_{lm} with the real spherical harmonics R_{lm} in expressions containing spherical harmonics.

3.4 Spatial Sampling

In practice, a continuous spherical pressure sensor is not available and therefore the sound field must be spatially sampled, such that the integral in (3.7) is replaced by a sum over a discrete number of microphones Q at positions $\mathbf{r}_q$, $q = 1, \ldots, Q$ [15, 18, 22]:

$$P_{lm}(\ell,\nu) \approx \sum_{q=1}^{Q} \mathcal{g}_{q,lm}\, P(\ell,\nu,\mathbf{r}_q). \tag{3.15}$$

In an approximation of a definite integral by a weighted sum, such as in (3.15), the *quadrature weights* $\mathcal{g}_{q,lm}$ are chosen such that the error involved in this approximation is minimized, and they are a function of the sampling configuration chosen. Error-free sampling is achieved when the approximation in (3.15) becomes an equality, or equivalently, when the *orthonormality error* [12] or *aliasing error* is zero [18]:

$$\sum_{q=1}^{Q} \mathcal{g}_{q,lm} Y_{l'm'}(\Omega_q) = \delta_{l,l'}\delta_{m,m'}, \tag{3.16}$$

where δ denotes the Kronecker delta defined in (2.19).

In the same way that a time domain signal must be temporally band-limited in order to be fully reconstructed from a finite number of samples without temporal aliasing, the SHD sound field must be order-limited ($P_{lm} = 0$ for $l > L_\text{f}$, where L_f is the order of the sound field) to be captured with a finite number of microphones without spatial aliasing [18]. A sound field that is limited to an order L_f is represented using a total of $\sum_{l=0}^{L_\text{f}} \sum_{m=-l}^{l} 1 = \sum_{l=0}^{L_\text{f}} (2l+1) = (L_\text{f}+1)^2$ eigenbeams, therefore all spatial sampling schemes require *at least* $(L_\text{f}+1)^2$ microphones to sample a sound field of order L_f without aliasing.

Depending on the sampling scheme, the number of microphones required $Q_\text{req}(L_\text{f})$ may be larger than $(L_\text{f}+1)^2$, as we will see in Sect. 3.4.1. It should be noted that the use of more than $(L_\text{f}+1)^2$ microphones is not necessarily 'wasteful', since additional microphones will reduce the level of sensor noise in the eigenbeams, even if they do not provide any additional spatial resolution.

Spatial aliasing occurs when high-order sound fields are captured using a spherical array with an insufficient number of sensors resulting in the high-order eigenbeams being aliased into the lower orders. A number of sampling schemes, three of which are presented below in Sect. 3.4.1, are aliasing-free (or have negligible aliasing) for order-limited functions. However, real sound fields are not actually order-limited: they are represented by an infinite series of spherical harmonics [23], and would therefore theoretically require an infinite number of microphones to be captured completely without spatial aliasing.

Nevertheless, the eigenbeams possess a property that reduces the spatial aliasing that results from the use of a finite number of microphones. Let us assume that we use a sampling scheme that is aliasing-free or has negligible aliasing for sound fields that are limited to an order L. This scheme requires a number of microphones $Q_\text{req}(L)$, where $Q_\text{req}(L) \geq (L+1)^2$ regardless of the sampling scheme. As we will see in Sect. 3.4.2, the magnitude of the eigenbeams decays rapidly for $l > kr$. We can therefore consider that when capturing a particular sound field, even if it is not limited to an order L, the aliasing error obtained will be negligible if $kr \ll L$

[18, 23], or equivalently, using the dispersion relation (2.4), if the operating frequency f satisfies $f \ll \frac{Lc}{2\pi r}$.

In this book, we will hereafter refer to L as the *array order*, and implicitly assume that the array under consideration employs a sampling scheme with $Q_{\text{req}}(L)$ microphones. The higher the array order, the greater the number of eigenbeams that can be acquired without (significant) spatial aliasing, and the higher the spatial resolution of the array.

Example: In order to acquire eigenbeams of order $L = 4$ using an array of radius $r = 4.2$ cm (the radius of the *Eigenmike* [16]) with negligible spatial aliasing, the operating frequency must be smaller than $f = \frac{Lc}{2\pi r} = 5.2$ kHz, and regardless of the positioning of the microphones, at least $(L+1)^2 = 25$ microphones will be required.

For higher operating frequencies, spatial anti-aliasing filters have been proposed to reduce the aliasing errors [2, 23]. The requirement that $kr \ll L$ implies that aliasing can also be avoided by increasing L with additional microphones or a more efficient sampling scheme, or by decreasing the radius of the array r. However, we will see in Sect. 3.4.2 that reducing the array radius leads to reduced robustness at low frequencies, and the choice of radius is therefore a compromise [14].

Since the number of microphones Q is usually much higher than the number of eigenbeams $(L+1)^2$, it can be advantageous to first transform the Q microphone signals to the SHD using real spherical harmonics, and then transform the $(L+1)^2$ eigenbeams to the STFT domain, as explained in Sect. 3.3. The computational complexity is reduced because the SHT is then real rather than complex, and only $(L+1)^2$ rather than Q STFTs need to be computed. In this case, the integral in (3.9) is replaced with the sum

$$p_{lm}[n] \approx \sum_{q=1}^{Q} \mathfrak{g}_{q,lm}\, p[n, \mathbf{r}_q]. \tag{3.17}$$

The quadrature weights $\mathfrak{g}_{q,lm}$ are then computed in the same way as for the complex SHT, except that the complex spherical harmonics Y_{lm} must be replaced with the real spherical harmonics R_{lm}.

In Sect. 3.4.1, we introduce various sampling schemes that can be used to sample the sound field with a spherical microphone array, and in Sect. 3.4.2, we introduce a number of commonly used spherical array configurations. Further discussion of array configurations is given in [19].

3.4.1 Sampling Schemes

The simplest sampling scheme is **equi-angle sampling**, where the inclination θ and azimuth ϕ are uniformly sampled at $2(L+1)$ angles given by $\theta_\iota = \frac{\pi\iota}{2L+2}$, $\iota = 0, \ldots, 2L+1$ and $\phi_j = \frac{2\pi j}{2L+2}$, $j = 0, \ldots, 2L+1$ [6, 18]. The scheme therefore requires a total of $Q_{\text{req}}(L) = 4(L+1)^2$ microphones. The quadrature weights are given by $\mathcal{G}_{q,lm} = \mathcal{G}_\iota Y_{lm}^*(\theta_\iota, \phi_j)$ [6, 18], where $q = j + \iota(2L+2) + 1$, and the term $\mathcal{G}_\iota$ compensates for the denser sampling in θ near the poles [6, 18]. The advantage of this scheme is the uniformity of the angle distributions, which can be useful when samples are taken by a rotating microphone; however, this comes at the expense of a relatively large number of required samples.

In **Gaussian sampling**, only half as many samples are needed: the azimuth is still sampled at $2(L+1)$ angles, whereas the inclination is sampled at only $L+1$ angles, requiring a total of $Q_{\text{req}}(L) = 2(L+1)^2$ microphones. The azimuth angles are the same as for equi-angle sampling, while the inclination angles must satisfy $\mathcal{P}_{L+1}(\cos\theta_\iota) = 0$, $\iota = 0, \ldots, L$ [18], where $\mathcal{P}_{L+1}$ is the Legendre polynomial of order $L+1$. The quadrature weights are then given by $\mathcal{G}_{q,lm} = \mathcal{G}_\iota Y_{lm}^*(\theta_\iota, \phi_j)$ [23], where $q = j + \iota(2L+2) + 1$ and the weights $\mathcal{G}_\iota$ are given in [3, 10]. The disadvantage of this scheme is that the inclination distribution is no longer uniform, meaning that if the microphones are mechanically rotated as in a scanning array [20], a fixed step size cannot be used [18]; however, for a fixed array configuration this is not likely to be a problem.

Finally, in **(nearly) uniform sampling**, the samples are (nearly) uniformly distributed on the sphere, in other words, the distance between each sample and its neighbours is (nearly) constant. A limited number of distributions perfectly satisfy this requirement, in which microphones are positioned at the centre of the faces or the vertices of the so-called platonic solids (the tetrahedron, cube, octahedron, dodecahedron, and icosahedron). However, there are nearly uniform distributions with negligible orthonormality error; for example, 32 microphones can be positioned at the centre of the faces of a truncated icosahedron [14]. The quadrature weights are given by $\mathcal{G}_{q,lm} = \frac{4\pi}{Q} Y_{lm}^*(\Omega_q)$ for uniform sampling [7, 27]. This sampling scheme requires a minimum of $(L+1)^2$ microphones; however, in contrast to the two previous schemes, the required number of microphones $Q_{\text{req}}(L)$ may be larger than $(L+1)^2$, depending on the chosen polyhedron.

Regardless of the sampling scheme, the Q quadrature weights associated with order l and degree m can be computed in the weighted least squares sense. First, (3.16) is expressed in matrix form as

$$\mathbf{Y}\,\boldsymbol{\mathcal{G}}_{lm} = \mathbf{e}_{lm}, \tag{3.18}$$

where $\mathbf{Y}$ is a matrix of size $(L+1)^2 \times Q$ that is defined as

$$\mathbf{Y} = \begin{bmatrix} Y_{00}(\Omega_1) & \cdots & Y_{00}(\Omega_Q) \\ Y_{1(-1)}(\Omega_1) & \cdots & Y_{1(-1)}(\Omega_Q) \\ Y_{10}(\Omega_1) & \cdots & Y_{10}(\Omega_Q) \\ Y_{11}(\Omega_1) & \cdots & Y_{11}(\Omega_Q) \\ Y_{2(-2)}(\Omega_1) & \cdots & Y_{2(-2)}(\Omega_Q) \\ \vdots & \ddots & \vdots \\ Y_{LL}(\Omega_1) & \cdots & Y_{LL}(\Omega_Q) \end{bmatrix}, \tag{3.19}$$

$\mathfrak{g}_{lm}$ is a vector of length Q containing the quadrature weights that is defined as

$$\mathfrak{g}_{lm} = [\mathfrak{g}_{1,lm}, \mathfrak{g}_{2,lm}, \cdots, \mathfrak{g}_{Q,lm}]^{\mathrm{T}}, \tag{3.20}$$

and $\mathbf{e}_{lm}$ is a vector of length $(L+1)^2$ where the $((l+1)\,l+m+1)$th element is one and the remaining elements are zero.

For $Q \geq (L+1)^2$, the quadrature weights for order l and degree m are then given in the weighted least squares sense by

$$\mathfrak{g}_{lm}^{\mathrm{WLS}} = \left(\mathbf{Y}^{\mathrm{H}}\mathbf{W}\mathbf{Y}\right)^{-1}\mathbf{Y}^{\mathrm{H}}\mathbf{W}\,\mathbf{e}_{lm}, \tag{3.21}$$

where $(\cdot)^{\mathrm{H}}$ denotes the Hermitian transpose, and $\mathbf{W}$ is a diagonal weighting matrix of size $(L+1)^2 \times (L+1)^2$. A common choice for the weighting matrix $\mathbf{W}$ is the identity matrix. Regularization techniques can be used to limit the white noise gain, and increase the robustness with respect to errors in the microphone positions.

In the rest of this book, uniform sampling will be employed, and it will be assumed that this sampling is aliasing-free. This is a reasonable assumption for arrays with a small radius (4–5 cm) and a few dozen microphones, operating at frequencies up to 4 kHz, as is typical in applications involving narrowband speech signals.

3.4.2 Array Configurations

The sound pressure captured by the microphones in a spherical array depends on the array properties, e.g., radius, configuration (open, rigid, dual-sphere, etc.), or microphone type. This dependence is captured by the frequency-dependent *mode strength* $b_l(k)$, which determines the amplitude of the lth-order eigenbeam(s) $P_{lm}(\ell, \nu)$ $(m = -l, \ldots, l)$. For a unit amplitude plane wave incident from a direction Ω_{s}, the SHD sound pressure and the mode strength $b_l(k)$ are related via the expression [14, 17, 26]

$$P_{lm}(\ell, \nu) = b_l(k) Y^*_{lm}(\Omega_s) \tag{3.22a}$$

$$= b_l\left(\frac{2\pi\nu f_s}{cK}\right) Y^*_{lm}(\Omega_s), \tag{3.22b}$$

where (3.2) is used to convert frequency indices ν to discrete values of the wavenumber k.

The simplest array configuration is the **open sphere** composed of omnidirectional microphones suspended in free space. It is assumed that the microphones and associated cabling and mounting brackets are acoustically transparent, that is to say, they have no effect on the measured sound field. In this case, the mode strength is given by [17, 26][1]

$$b_l(k) = i^l j_l(kr), \tag{3.23}$$

where $j_l(kr)$ is the spherical Bessel function of order l. This equation can be derived by applying the SHT to the SHE of the expression for a plane wave in (2.29). The open configuration is convenient for large array radii, where a rigid array would be impractical, and for scanning arrays [20].

When processing the eigenbeams captured using the spherical array, it is necessary to remove the dependence on the array properties by dividing the eigenbeams by $b_l(k)$, thereby removing the frequency-dependence of the eigenbeams. This process is often referred to as *mode strength compensation*. The open sphere mode strength is plotted in Fig. 3.3 (dashed line); it can be seen that there are zeros at certain frequencies (for certain values of kr). As a result, the open array may suffer from poor robustness at these frequencies, where measurement noise will be significantly amplified. In addition, it can be seen that for $l > 0$, at low frequencies the mode strength is very small; as a result, high-order eigenbeams are generally not used at low frequencies [13].

The **rigid sphere** is a popular alternative to the open sphere. In this configuration, omnidirectional microphones are mounted on a rigid spherical baffle, and the array is therefore no longer acoustically transparent: the sound waves are scattered by the sphere. An example of a rigid spherical array, the *Eigenmike* [16], is shown in Fig. 1.2. The mode strength for a rigid sphere of radius r_a is given by [14, 17][2]

$$b_l(kr_a, kr) = i^l \left(j_l(kr) - \frac{j_l'(kr_a)}{h_l^{(2)'}(kr_a)} h_l^{(2)}(kr) \right), \tag{3.24}$$

[1]This equation assumes the sign convention used in engineering. If the acoustics convention is used, this equation is given by $b_l(k) = (-i)^l j_l(kr)$. For more information on the effects of the choice of sign convention, see Sect. 2.3.

[2]This equation assumes the sign convention used in engineering. If the acoustics convention is used, this equation is given by $b_l(kr_a, kr) = (-i)^l \left(j_l(kr) - \frac{j_l'(kr_a)}{h_l^{(1)'}(kr_a)} h_l^{(1)}(kr) \right)$, where $h_l^{(1)} = \left(h_l^{(2)}\right)^*$ denotes the spherical Hankel function of the first kind. For more information on the effects of the choice of sign convention, see Sect. 2.3.

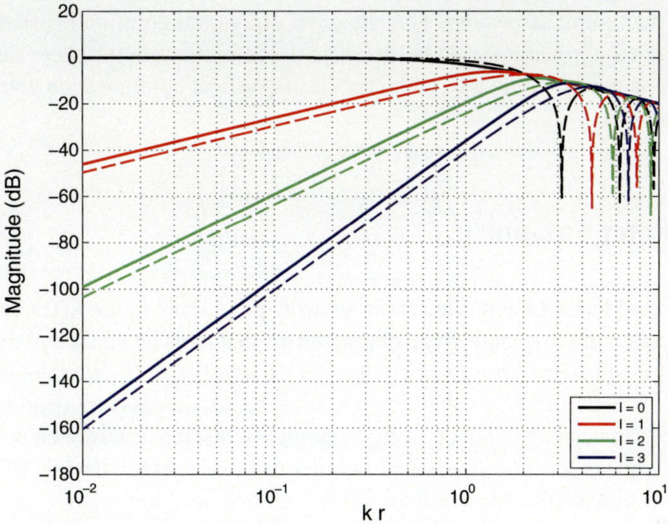

Fig. 3.3 Magnitude of the mode strength $b_l(k)$ for orders $l \in \{0, 1, 2, 3\}$ as a function of kr. The *solid lines* denote a rigid sphere, and the *dashed lines* denote an open sphere. Copyright © Daniel Jarrett. Used with permission

where j_l' and $h_l^{(2)'}$ respectively denote the first derivatives of j_l and $h_l^{(2)}$ with respect to the argument, and $h_l^{(2)}$ is the spherical Hankel function of the second kind. As is also the case in the example of the Eigenmike, the microphones are normally positioned on the surface of the rigid sphere (i.e., $r = r_\mathrm{a}$), therefore we define $b_l(k) = b_l(kr, kr)$. The second term in (3.24) compared to (3.23) accounts for the effect of scattering.

From the plot of the rigid sphere mode strength in Fig. 3.3 (solid line), an advantage of the rigid sphere can be observed: it does not suffer from zeros in its mode strength at any frequency, unlike the open sphere. In addition, the scattering effects of the rigid sphere can be calculated precisely and can be incorporated into the eigenbeam processing framework. For a detailed discussion of the scattering effects of the rigid sphere, the reader is referred to Sect. 4.2.3.

The rigid array configuration will be used for most of the work in this book. A number of other configurations have been proposed, and will be mentioned briefly in the following. The mode strength expressions for the following configurations can be found in [21] and the references therein. The hemisphere [11] exploits the symmetry of the sound field by mounting the array on a rigid surface. The open dual-sphere [4], comprised of two spheres with different radii, and the open sphere with cardioid microphones [4] both overcome the problem of zeros in the open sphere mode strength, although cardioid microphones are not as readily available as omnidirectional microphones. Finally, in the free sampling configuration the microphones can be placed anywhere on the surface of a rigid sphere [12]; their positions are then optimized to robustly achieve an optimal approximation of a desired beam pattern, or maximum directivity. The choice of array configuration is usually based on the intended application; for example, in a conference room where the microphone array is placed on a large table, the hemispherical configuration could be the most appropriate.

3.5 Chapter Summary

This chapter addressed the preliminaries of array processing in the SHD. In particular, it introduced the two mathematical transforms that must be applied to the signals acquired by a spherical array, namely, the STFT and the SHT. We then explained how the SHT, which involves the integration of the sound pressure over a continuous surface, can be approximated using a number of discrete microphones, and how these microphones must be positioned in order to avoid spatial aliasing. Finally, we discussed a number of array configurations, and explored the relative strengths of the two most common configurations: the open and rigid spheres with omnidirectional microphones.

References

1. Allen, J., Radiner, L.: A unified approach to short-time Fourier analysis and synthesis. Proc. IEEE **65**(11), 1558–1564 (1977)
2. Alon, D.L., Rafaely, B.: Beamforming with optimal aliasing cancellation in spherical microphone arrays. IEEE/ACM Trans. Audio Speech Lang. Process. **24**(1), 196–210 (2016)
3. Arfken, G.B., Weber, H.J.: Mathematical Methods for Physicists, 5th edn. Academic Press, San Diego (2001)
4. Balmages, I., Rafaely, B.: Open-sphere designs for spherical microphone arrays. IEEE Trans. Audio Speech Lang. Process. **15**(2), 727–732 (2007). doi:10.1109/TASL.2006.881671
5. Brookes, D.M.: VOICEBOX: a speech processing toolbox for MATLAB. http://www.ee.ic.ac.uk/hp/staff/dmb/voicebox/voicebox.html (1997–2013)
6. Driscoll, J.R., Healy, D.M.: Computing Fourier transforms and convolutions on the 2-sphere. Adv. Appl. Math. **15**(2), 202–250 (1994). doi:10.1006/aama.1994.1008
7. Elko, G.W., Meyer, J.: Spherical microphone arrays for 3D sound recordings. In: Huang, Y., Benesty, J. (eds.) Audio Signal Processing for Next-Generation Multimedia Communication Systems, pp. 67–89 (2004)
8. Jarrett, D.P.: Spherical microphone array processing for acoustic parameter estimation and signal enhancement. Ph.D. thesis, Imperial College London (2013)
9. Kennedy, R., Sadeghi, P.: Hilbert Space Methods. Cambridge University Press, Cambridge (2013)
10. Krylov, V.I.: Approximate Calculation of Integrals. MacMillan, New York (1962)
11. Li, Z., Duraiswami, R.: Hemispherical microphone arrays for sound capture and beamforming. In: Proceedings of the IEEE Workshop on Applications of Signal Processing to Audio and Acoustics, pp. 106–109 (2005)
12. Li, Z., Duraiswami, R.: Flexible and optimal design of spherical microphone arrays for beamforming. IEEE Trans. Audio Speech Lang. Process. **15**(2), 702–714 (2007). doi:10.1109/TASL.2006.876764

13. Meyer, J., Agnello, T.: Spherical microphone array for spatial sound recording. In: Proceedings of the Audio Engineering Society Convention, pp. 1–9. New York, NY, USA (2003)
14. Meyer, J., Elko, G.: A highly scalable spherical microphone array based on an orthonormal decomposition of the soundfield. In: Proceedings of the IEEE International Conference on Acoustics, Speech, and Signal Processing (ICASSP), vol. 2, pp. 1781–1784 (2002)
15. Meyer, J., Elko, G.W.: Position independent close-talking microphone. Signal Process. **86**(6), 1254–1259 (2006). doi:10.1016/j.sigpro.2005.05.036
16. MH Acoustics LLC: The Eigenmike microphone array. http://www.mhacoustics.com/mh_acoustics/Eigenmike_microphone_array.html
17. Rafaely, B.: Plane-wave decomposition of the pressure on a sphere by spherical convolution. J. Acoust. Soc. Am. **116**(4), 2149–2157 (2004)
18. Rafaely, B.: Analysis and design of spherical microphone arrays. IEEE Trans. Speech Audio Process. **13**(1), 135–143 (2005). doi:10.1109/TSA.2004.839244
19. Rafaely, B.: Fundamentals of Spherical Array Processing. Springer, Berlin (2015)
20. Rafaely, B., Balmages, I., Eger, L.: High-resolution plane-wave decomposition in an auditorium using a dual-radius scanning spherical microphone array. J. Acoust. Soc. Am. **122**(5), 2661–2668 (2007)
21. Rafaely, B., Kleider, M.: Spherical microphone array beam steering using Wigner-D weighting. IEEE Signal Process. Lett. **15**, 417–420 (2008). doi:10.1109/LSP.2008.922288
22. Rafaely, B., Peled, Y., Agmon, M., Khaykin, D., Fisher, E.: Spherical microphone array beamforming. In: Cohen I., Benesty J., Gannot S. (eds.) Speech Processing in Modern Communication: Challenges and Perspectives, vol. 11. Springer, Heidelberg (2010)
23. Rafaely, B., Weiss, B., Bachmat, E.: Spatial aliasing in spherical microphone arrays. IEEE Trans. Signal Process. **55**(3), 1003–1010 (2007). doi:10.1109/TSP.2006.888896
24. Rickard, S., Yilmaz, Z.: On the approximate W-disjoint orthogonality of speech. In: Proceedings of the IEEE International Conference on Acoustics, Speech and Signal Processing (ICASSP), vol. 1, pp. 529–532 (2002)
25. Smith, J.: Spectral Audio Signal Processing. W3K Publishing (2011)
26. Teutsch, H.: Wavefield decomposition using microphone arrays and its application to acoustic scene analysis. Ph.D. thesis, Friedrich-Alexander Universität Erlangen-Nürnberg (2005)
27. Williams, E.G.: Fourier Acoustics: Sound Radiation and Nearfield Acoustical Holography, 1st edn. Academic Press, London (1999)

Chapter 4
Spherical Array Acoustic Impulse Response Simulation

In general, the evaluation of acoustic signal processing algorithms, such as direction of arrival (DOA) estimation (see Chap. 5) and speech enhancement (see Chap. 9) algorithms, makes use of simulated acoustic transfer functions (ATFs). By using simulated ATF it is possible to evaluate comprehensively an algorithm under many acoustic conditions, such as a range of reverberation times, room dimensions and source-array distances. Allen and Berkley's image method [2] is a widely used approach to simulate ATFs between an omnidirectional sound source and one or more microphones in a reverberant environment. In the last few decades, several extensions have been proposed [21, 29].

In recent years the use of spherical microphone arrays has become prevalent. These arrays are commonly of one of two types (discussed in Sect. 3.4.2): the open array, where microphones are suspended in free space on an 'open' sphere, and the rigid array, where microphones are mounted on a rigid baffle. As discussed in the previous chapter, the rigid sphere is often preferred as it improves the numerical stability of many processing algorithms [32] and its acoustic scattering effects are can be calculated precisely [25].

Currently, many works relating to spherical array processing consider only free-field responses; however, when a rigid array is used, the rigid baffle causes scattering of the sound waves incident upon the array that the image method does not consider. This scattering has an effect on the ATFs, especially at high frequencies and/or for microphones situated on the occluded side of the array. Furthermore the reverberation due to room boundaries such as walls, ceiling and floor must also be considered, particularly in small rooms or rooms with strongly reflective surfaces.

Portions of this chapter were first published in the *Journal of the Acoustical Society of America* [17], and are reproduced in accordance with the Acoustical Society of America's Transfer of Copyright Agreement. The content of [17] has been edited here for brevity and to standardize the notation.

D. Jarrett et al., *Theory and Applications of Spherical Microphone Array Processing*, Springer Topics in Signal Processing 9,
DOI 10.1007/978-3-319-42211-4_4

While measured transfer functions include both these effects, they are both time-consuming and expensive to acquire over a wide range of geometries and rooms. A method for simulating ATFs in a reverberant room while accounting for scattering is therefore essential, allowing for fast, comprehensive and repeatable testing. In this chapter, we present the SMIR (Spherical Microphone array Impulse Response) method that combines a model of the scattering in the spherical harmonic domain (SHD) with a version of the image method that accounts for reverberation in a computationally efficient way [16, 17].

The simulated ATFs include the direct path, reflections due to room reverberation, scattering of the direct path and scattering of the reverberant reflections. Reflections of the scattered sound and multiple interactions between the room boundaries and the sphere are excluded as they do not contribute significantly to the sound field, provided the distances between the room boundaries and the sphere are several times the sphere's radius [11], which is easily achieved in the case of a small scatterer [4]. Furthermore, we assume an empty rectangular shoebox room (with the exception of the rigid sphere) and specular reflections, as was assumed in the conventional image method [2]. Finally, the scattering model used assumes a perfectly rigid baffle, without absorption.

In this chapter, we first briefly summarize Allen and Berkley's image method and then present the SMIR method in the SHD. Next, we discuss some implementation aspects, namely the truncation of an infinite sum in the ATF expression and the reduction of the method's computational complexity, and then provide a pseudocode description of the method. An open-source software implementation is available online [14]. Finally, we show some example uses of the method and, where possible, compare the simulated results obtained with theoretical models.

4.1 Allen and Berkley's Image Method

The source-image or image method [2] is one of the most commonly used room acoustics simulation methods in the acoustic signal processing community. The principle of the method is to model an ATF as the sum of a direct path component and a number of discrete reflections, each of these components being represented in the ATF by a free-space Green's function. In this section, we review the free-space Green's function and the image method.

4.1.1 Green's Function

As detailed in Sect. 2.1, for a source at a position ${}_{\llcorner}\mathbf{r}_{\mathrm{s}}$ and a receiver at a position ${}_{\llcorner}\mathbf{r}$,[1] the free-space Green's function, a solution to the inhomogeneous Helmholtz equation

---

[1] Vectors in Cartesian coordinates are denoted with a corner mark $\llcorner$ to distinguish them from vectors in spherical coordinates, which are used throughout this book and will be introduced later in the chapter.

applying the Sommerfeld radiation condition, is given by[2]

$$G(\llcorner\mathbf{r}|\llcorner\mathbf{r}_s, k) = \frac{e^{-ik||\llcorner\mathbf{r}-\llcorner\mathbf{r}_s||}}{4\pi\,||\llcorner\mathbf{r}-\llcorner\mathbf{r}_s||}, \tag{4.1}$$

where $||\cdot||$ denotes the 2-norm and the wavenumber k is related to frequency f (in Hz), angular frequency ω (in rad $\cdot$ s^{-1}) and the speed of sound c (in m $\cdot$ s^{-1}) via the dispersion relation $k = \omega/c = 2\pi f/c$.

In the time-domain, the Green's function is given by

$$g(\llcorner\mathbf{r}|\llcorner\mathbf{r}_s, t) = \frac{\delta\left(t - \frac{||\llcorner\mathbf{r}-\llcorner\mathbf{r}_s||}{c}\right)}{4\pi\,||\llcorner\mathbf{r}-\llcorner\mathbf{r}_s||}, \tag{4.2}$$

where δ is the Dirac delta function and t is time. This corresponds to a pure impulse at time $t = \frac{||\llcorner\mathbf{r}-\llcorner\mathbf{r}_s||}{c}$, the propagation time from $\llcorner\mathbf{r}_s$ to $\llcorner\mathbf{r}$.

4.1.2 Image Method

Consider a rectangular room with length L_x, width L_y and height L_z. The reflection coefficients of the four walls, floor and ceiling are β_{x_1}, β_{x_2}, β_{y_1}, β_{y_2}, β_{z_1} and β_{z_2}, where the a_1 coefficients ($a \in \{x, y, z\}$) correspond to the boundaries at $a = 0$ and the a_2 coefficients correspond to the boundaries at $a = L_a$.

If the sound source is located at $\llcorner\mathbf{r}_s = (x_s, y_s, z_s)$ and the receiver is located at $\llcorner\mathbf{r} = (x, y, z)$, the images obtained using the walls at $x = 0$, $y = 0$ and $z = 0$ can be expressed as a vector $\llcorner\mathbf{R_p}$:

$$\llcorner\mathbf{R_p} = [x_s - x + 2p_x x,\ y_s - y + 2p_y y,\ z_s - z + 2p_z z], \tag{4.3}$$

where each of the elements in $\mathbf{p} = (p_x, p_y, p_z)$ can take values 0 or 1, thus resulting in eight combinations that form a set $\mathcal{P}$. To consider all reflections we also define a vector $\llcorner\mathbf{R_m}$ which we add to $\llcorner\mathbf{R_p}$:

$$\llcorner\mathbf{R_m} = [2m_x L_x, 2m_y L_y, 2m_z L_z], \tag{4.4}$$

where each of the elements in $\mathbf{m} = (m_x, m_y, m_z)$ can take values between $-N_m$ and N_m, and N_m is used to limit computational complexity and circular convolution errors, thus resulting in a set $\mathcal{M}$ of $(2N_m + 1)^3$ combinations. The image positions in the x and y dimensions are illustrated in Fig. 4.1.

[2] This expression assumes the sign convention commonly used in electrical engineering, whereby the temporal Fourier transform is defined as $\mathcal{F}(\omega) = \int_{-\infty}^{\infty} f(t)e^{-i\omega t}dt$. For more information on this sign convention, the reader is referred to Sect. 2.3.

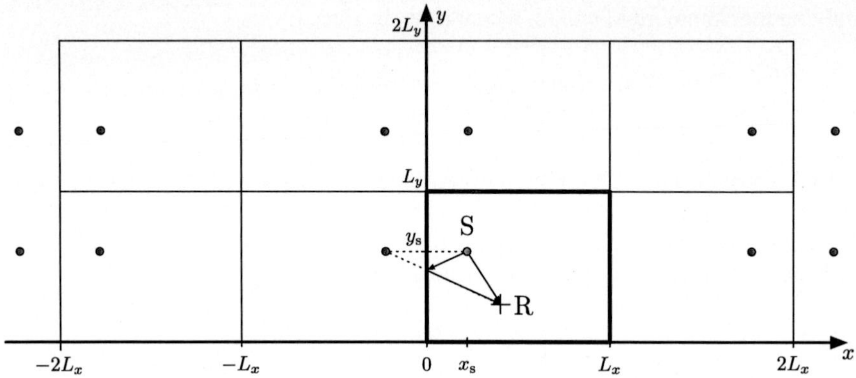

Fig. 4.1 A slice through the image space showing the positions of the images in the x and y dimensions, with a source S and receiver R. The full image space has three dimensions (x, y and z). An example of a reflected path (first-order reflection about the x-axis) is also shown

The distance between an image and the receiver is given by $\left|\left|\llcorner\mathbf{R_p} + \llcorner\mathbf{R_m}\right|\right|$. Using (4.1), the ATF H is then given by

$$H(\llcorner\mathbf{r}|\llcorner\mathbf{r}_s, k) = \sum_{\mathbf{p}\in\mathcal{P}} \sum_{\mathbf{m}\in\mathcal{M}} \beta_{x_1}^{|m_x+p_x|} \beta_{x_2}^{|m_x|} \beta_{y_1}^{|m_y+p_y|} \beta_{y_2}^{|m_y|} \beta_{z_1}^{|m_z+p_z|} \beta_{z_2}^{|m_z|} \times \frac{e^{-ik\left|\left|\llcorner\mathbf{R_p}+\llcorner\mathbf{R_m}\right|\right|}}{4\pi\left|\left|\llcorner\mathbf{R_p} + \llcorner\mathbf{R_m}\right|\right|}. \tag{4.5}$$

Using (4.2), we obtain the acoustic impulse response (AIR)

$$h(\llcorner\mathbf{r}|\llcorner\mathbf{r}_s, t) = \sum_{\mathbf{p}\in\mathcal{P}} \sum_{\mathbf{m}\in\mathcal{M}} \beta_{x_1}^{|m_x+p_x|} \beta_{x_2}^{|m_x|} \beta_{y_1}^{|m_y+p_y|} \beta_{y_2}^{|m_y|} \beta_{z_1}^{|m_z+p_z|} \beta_{z_2}^{|m_z|} \times \frac{\delta\left(t - \frac{\left|\left|\llcorner\mathbf{R_p}+\llcorner\mathbf{R_m}\right|\right|}{c}\right)}{4\pi\left|\left|\llcorner\mathbf{R_p} + \llcorner\mathbf{R_m}\right|\right|}. \tag{4.6}$$

4.2 SMIR Method in the Spherical Harmonic Domain

There exists a compact analytical expression for the scattering due to the rigid sphere in the SHD, therefore we first express the free-space Green's function in this domain, and then use this to form an expression for the ATF including scattering.

4.2.1 Green's Function

We define position vectors in spherical coordinates relative to the centre of our array. Letting r be the array radius and Ω an inclination-azimuth pair, the microphone position vector is defined as $\mathbf{r} \triangleq (r, \Omega)$ where $\Omega = (\theta, \phi)$. Similarly, the source position vector is given by $\mathbf{r}_s \triangleq (r_s, \Omega_s)$ where $\Omega_s = (\theta_s, \phi_s)$. Consistent with our approach in previous chapters, it is hereafter assumed that where the addition, 2-norm or scalar product operations are applied to spherical polar vectors, they have previously been converted to Cartesian coordinates using (2.12). In addition, we assume that the source is outside the array, i.e., $r_s > r$.

The free-space Green's function (4.1) can be expressed in the SHD using the spherical harmonic expansion (SHE) in (2.22) [40]:

$$\begin{aligned} G(\mathbf{r}|\mathbf{r}_s, k) &= \frac{e^{-ik||\mathbf{r}-\mathbf{r}_s||}}{4\pi\,||\mathbf{r}-\mathbf{r}_s||} \\ &= -ik \sum_{l=0}^{\infty} \sum_{m=-l}^{l} j_l(kr) h_l^{(2)}(kr_s) Y_{lm}^*(\Omega_s) Y_{lm}(\Omega) \end{aligned} \tag{4.7}$$

where Y_{lm} is the spherical harmonic function of order l and degree m, j_l is the spherical Bessel function of order l and $h_l^{(2)}$ is the spherical Hankel function of the second kind and of order l. This decomposition is also known as a spherical Fourier series expansion or spherical harmonic decomposition of the Green's function.

Using the spherical harmonic addition theorem (2.23), which in many cases reduces the complexity of the implementation, we can simplify the Green's function in (4.7) to

$$G(\mathbf{r}|\mathbf{r}_s, k) = \frac{-ik}{4\pi} \sum_{l=0}^{\infty} j_l(kr) h_l^{(2)}(kr_s)(2l+1)\mathcal{P}_l(\cos\Theta_{\mathbf{r},\mathbf{r}_s}), \tag{4.8}$$

where $\mathcal{P}_l$ is the Legendre polynomial of order l and $\Theta_{\mathbf{r},\mathbf{r}_s}$ is the angle between $\mathbf{r}$ and $\mathbf{r}_s$. The cosine of the angle $\Theta_{\mathbf{r},\mathbf{r}_s}$ is obtained as the dot product of the two normalized vectors $\hat{\mathbf{r}}_s = \mathbf{r}_s / r_s$ and $\hat{\mathbf{r}} = \mathbf{r}/r$:

$$\cos\Theta_{\mathbf{r},\mathbf{r}_s} = \hat{\mathbf{r}} \cdot \hat{\mathbf{r}}_s \tag{4.9a}$$

$$= \sin\theta\cos\phi\sin\theta_s\cos\phi_s + \sin\theta\sin\phi\sin\theta_s\sin\phi_s + \cos\theta\cos\theta_s \tag{4.9b}$$

$$= \sin\theta\sin\theta_s\cos(\phi - \phi_s) + \cos\theta\cos\theta_s. \tag{4.9c}$$

4.2.2 Neumann Green's Function

The free-space Green's function describes the propagation of sound in free space only. However, when a rigid sphere is present, a boundary condition must hold: the radial velocity must vanish on the surface of the sphere. The function $G_{\mathrm{N}}(\mathbf{r}|\mathbf{r}_{\mathrm{s}}, k)$ satisfying this boundary condition is called the *Neumann Green's* function, and describes the sound propagation between a point $\mathbf{r}_{\mathrm{s}}$ and a point $\mathbf{r}$ on the rigid sphere [40]:

$$\begin{aligned} G_{\mathrm{N}}(\mathbf{r}|\mathbf{r}_{\mathrm{s}}, k) &= G(\mathbf{r}|\mathbf{r}_{\mathrm{s}}, k) - \frac{-ik}{4\pi}\sum_{l=0}^{\infty}\frac{j_l'(kr)h_l^{(2)}(kr)}{h_l^{(2)'}(kr)}h_l^{(2)}(kr_{\mathrm{s}})(2l+1)\mathcal{P}_l(\cos\Theta_{\mathbf{r},\mathbf{r}_{\mathrm{s}}}) \\ &= \frac{-ik}{4\pi}\sum_{l=0}^{\infty} i^{-l} b_l(k) h_l^{(2)}(kr_{\mathrm{s}})(2l+1)\mathcal{P}_l(\cos\Theta_{\mathbf{r},\mathbf{r}_{\mathrm{s}}}), \end{aligned} \quad (4.10)$$

where $(\cdot)'$ denotes the first derivative and the term

$$b_l(k) = i^l\left(j_l(kr) - \frac{j_l'(kr)}{h_l^{(2)'}(kr)}h_l^{(2)}(kr)\right) \quad (4.11)$$

is often called the (farfield) *mode strength*. The Wronskian relation [40, Eq. 6.67]

$$j_l(x)h_l^{(2)'}(x) - j_l'(x)h_l^{(2)}(x) = -\frac{i}{x^2} \quad (4.12)$$

allows us to simplify (4.11) to

$$b_l(k) = \frac{-i^{l+1}}{h_l^{(2)'}(kr)\,(kr)^2}. \quad (4.13)$$

For the open sphere, substituting $b_l(k) = i^l j_l(kr)$ into (4.10) yields the free-space Green's function $G(\mathbf{r}|\mathbf{r}_{\mathrm{s}}, k)$.

4.2.3 Scattering Model

The rigid sphere scattering model[3] used by the SMIR method has a long history in the literature; it was first developed by Clebsch and Rayleigh in 1871–72 [23]. It is

[3] Some texts [9] refer to the scattering effect as diffraction, although Morse and Ingard note that *"When the scattering object is large compared with the wavelength of the scattered sound, we usually say the sound is reflected and diffracted, rather than scattered"* [28], therefore in the case of spherical microphone arrays (particularly rigid ones which tend to be relatively small for practical reasons), scattering is possibly the more appropriate term.

presented in a number of acoustics texts [28, 36, 40], and is used in many theoretical analyses for spherical microphone arrays [26, 33].

4.2.3.1 Theoretical Behaviour

The behaviour of the scattering model is illustrated in Fig. 4.2, which plots the magnitude of the response between a source and a receiver on a rigid sphere of radius 5 cm for a source-array distance of 1 m, as a function of frequency and DOA. The response was normalized using the free-field/open sphere response, therefore the plot shows only the effect due to scattering. Due to rotational symmetry, we only looked at the one-dimensional DOA, instead of looking at both azimuth and inclination, and limited the DOA to the 0–180° range.

When the source is located on the same side of the sphere as the receiver and the direction of arrival is 0°, the rigid sphere response is greater than the open sphere response due to constructive scattering, tending towards a 6 dB magnitude gain compared to the open sphere at infinite frequency. The response on the back side of the rigid sphere is generally lower than in the open sphere case and lower than on the front side, as one would intuitively expect due to it being occluded. However, at the very back of the sphere, when the DOA is 180°, we observe a narrow *bright spot*: the waves propagating around the sphere all arrive in phase at the 180° point and as a result sum constructively.

The polar plot of the magnitude response is shown in Fig. 4.3 and illustrates both the amplification on the front side of the sphere, and attenuation on the back side of

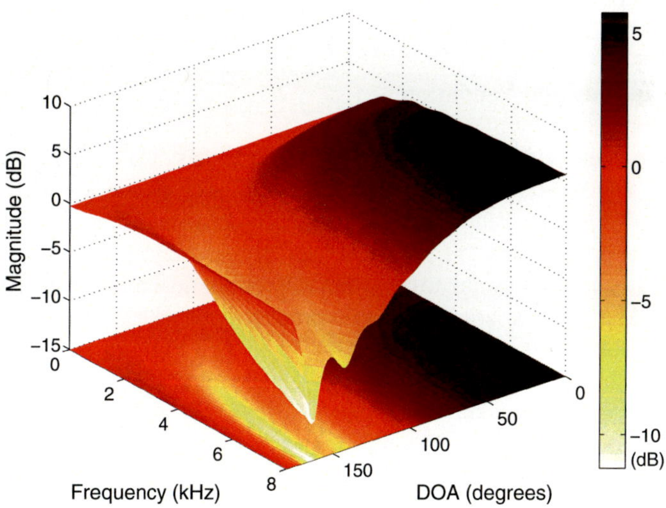

Fig. 4.2 Magnitude of the response between a source and a receiver placed on a rigid sphere of radius 5 cm at a distance of 1 m, as a function of frequency and DOA. The response was normalized with respect to the free-field response

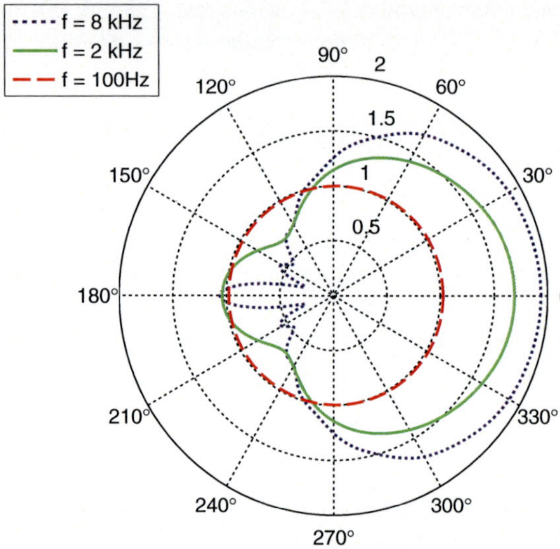

Fig. 4.3 Polar plot of the magnitude of the response between a source and a receiver placed on a rigid sphere of radius 5 cm, at a distance of 1 m, for various frequencies

the sphere, which both increase with increasing frequency. It should be noted that although the above results are for a fixed sphere radius, as the scattering model is a function of kr, the effects of a change in radius are the same as a change in frequency; indeed the relevant factor is the radius of the sphere relative to the wavelength.

These substantial differences between the open and rigid sphere responses confirm the need for a simulation method which accounts for scattering, even for sphere radii as small as 5 cm.

4.2.3.2 Experimental Validation

In addition to being widely used in theory, this model has also been experimentally validated by Duda and Martens [9] using a single microphone inserted in a hole drilled through a 10.9 cm radius bowling ball placed in an anechoic chamber. This is a reasonable approximation to a spherical microphone array; indeed a bowling ball was used by Li and Duraiswami to construct a hemispherical microphone array [22].

Duda and Martens's experimental results broadly agree with the theoretical model. In our case we are most interested in the results in their Fig. 12a where the source-array distance is largest (20 times the array radius), as in typical spherical array usage scenarios the source is unlikely to be much closer to the array than this. The only notable difference between the theoretical and experimental results in this case is for a direction of arrival of 180°, where the high frequency response is found to be lower than expected. The authors suggest this is due to small alignment errors, which would indeed have an effect given the narrowness of the bright spot in the model (see Fig. 4.3 for $f = 8$ kHz). Given these results, we conclude that the use of this scattering model is sufficiently accurate for simulating a small rigid array, such as the *Eigenmike* [27].

4.2.4 SMIR Method

We now present the SMIR method proposed in [16, 17], incorporating the SHE presented in Sect. 4.2.1 and the scattering model introduced in Sect. 4.2.2.

Due to the differences between the SHD Neumann Green's function in (4.10) and the spatial domain Green's function in (4.1), as well as the directionality of the array's response, two changes must be made to the image position vectors ${}_{\llcorner}\mathbf{R}_{\mathbf{p}}$ and ${}_{\llcorner}\mathbf{R}_{\mathbf{m}}$ in the SMIR method. Firstly, to compute the SHE in the Neumann Green's function, we require the distance between each image and the *centre* of the array [r_{s} in (4.10)]; this is accomplished by computing the image position vectors using the position of the centre of the array rather than the position of the receiver. Secondly, to compute the SHE we require the angle between each image and the receiver with respect to the centre of the array [$\Theta_{\mathbf{r},\mathbf{r}_{\mathrm{s}}}$ in (4.10)]. In Allen and Berkley's image method, the direction of the vector ${}_{\llcorner}\mathbf{R}_{\mathbf{p}} + {}_{\llcorner}\mathbf{R}_{\mathbf{m}}$ is not always the same: in some cases it points from the receiver to the image and in others it points from the image to the receiver. This is not an issue for the image method as only the norm of this vector is used. Because we also require the angle of the images in the SMIR method, we modify the definition of ${}_{\llcorner}\mathbf{R}_{\mathbf{p}}$ such that the vector ${}_{\llcorner}\mathbf{R}_{\mathbf{p}} + {}_{\llcorner}\mathbf{R}_{\mathbf{m}}$ always points from the centre of the array to the image.

We now incorporate these two changes into the definition of the image vectors ${}_{\llcorner}\mathbf{R}_{\mathbf{p}}$ and ${}_{\llcorner}\mathbf{R}_{\mathbf{m}}$. If the sound source is located at ${}_{\llcorner}\mathbf{r}_{\mathrm{s}} = (x_{\mathrm{s}}, y_{\mathrm{s}}, z_{\mathrm{s}})$ and the centre of the sphere is located at ${}_{\llcorner}\mathbf{r}_{\mathrm{a}} = (x_{\mathrm{a}}, y_{\mathrm{a}}, z_{\mathrm{a}})$, the images obtained using the walls at $x = 0$, $y = 0$ and $z = 0$ are expressed as a vector ${}_{\llcorner}\mathbf{R}_{\mathbf{p}}$:

$$
{}_{\llcorner}\mathbf{R}_{\mathbf{p}} = [x_{\mathrm{s}} - 2p_x x_{\mathrm{s}} - x_{\mathrm{a}}, y_{\mathrm{s}} - 2p_y y_{\mathrm{s}} - y_{\mathrm{a}}, z_{\mathrm{s}} - 2p_z z_{\mathrm{s}} - z_{\mathrm{a}}]. \tag{4.14}
$$

For brevity we define ${}_{\llcorner}\mathbf{R}_{\mathbf{p},\mathbf{m}} \triangleq {}_{\llcorner}\mathbf{R}_{\mathbf{p}} + {}_{\llcorner}\mathbf{R}_{\mathbf{m}}$, allowing us to express the distance between an image and the centre of the sphere as $||{}_{\llcorner}\mathbf{R}_{\mathbf{p},\mathbf{m}}||$ and the angle between the image and the receiver as $\Theta_{\mathbf{r},\mathbf{R}_{\mathbf{p},\mathbf{m}}}$, computed in the same way as (4.9), where $\mathbf{R}_{\mathbf{p},\mathbf{m}}$ denotes the image positions in spherical coordinates. The image positions in the x dimension are illustrated in Fig. 4.4. Finally, the ATF $H(\mathbf{r}|\mathbf{r}_{\mathrm{s}}, k)$ is the weighted sum of the individual responses $G_{\mathrm{N}}(\mathbf{r}|\mathbf{R}_{\mathbf{p},\mathbf{m}}, k)$ for each of the images[4]

$$
\begin{aligned}
H(\mathbf{r}|\mathbf{r}_{\mathrm{s}}, k) = \sum_{\mathbf{p}\in\mathcal{P}} \sum_{\mathbf{m}\in\mathcal{M}} & \beta_{x_1}^{|m_x - p_x|} \beta_{x_2}^{|m_x|} \beta_{y_1}^{|m_y - p_y|} \beta_{y_2}^{|m_y|} \beta_{z_1}^{|m_z - p_z|} \beta_{z_2}^{|m_z|} \\
& \times G_{\mathrm{N}}(\mathbf{r}|\mathbf{R}_{\mathbf{p},\mathbf{m}}, k).
\end{aligned} \tag{4.15}
$$

[4]The sign in the powers of β is different from that in Allen and Berkley's conventional image method, due to the change in the definition of ${}_{\llcorner}\mathbf{R}_{\mathbf{p}}$ that is required for the SMIR method.

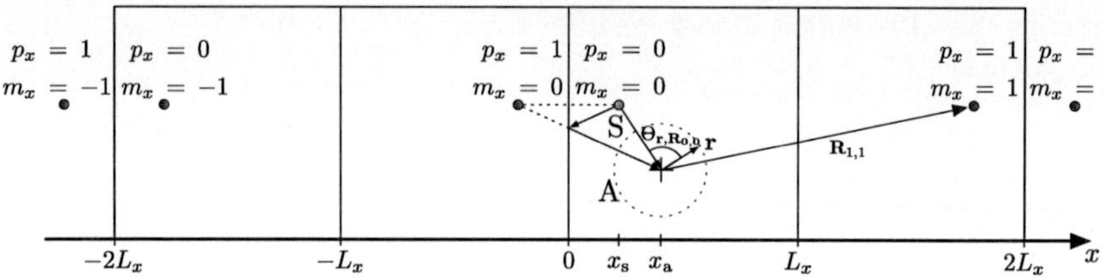

Fig. 4.4 A slice through the image space showing the positions of the images in the x dimension, with a source S and array A. The full image space has three dimensions (x, y and z). An example of a reflected path is shown for the image with $p_x = 1$ and $m_x = 0$

Since we are working in the wavenumber domain, we can allow for frequency dependent boundary reflection coefficients in (4.15), if desired. The reflection coefficients would then be written as $\beta_{x_1}(k)$, $\beta_{x_2}(k)$ and so on. Chen and Maher [7] provide some measured reflection coefficients for a wall, window, floor and ceiling.

4.3 Implementation

4.3.1 Truncation Error

To compute the expression for the ATF in (4.15), the sum over an infinite number of orders l in the Neumann Green's function G_{N} must be approximated by a sum $\hat{G}_{\text{N}}$ over a finite order L. Choosing L too small will result in a large approximation error, while choosing L too large will result in too high a computational complexity. We now investigate the approximation error in order to provide some guidelines for the choice of the order L. The results for an open sphere are provided for reference, and were computed by using a truncated SHE of the Green's function $\hat{G}$ instead of a Neumann Green's function.

For an open sphere, the error can be determined exactly because the Green's function is a decomposition of the closed-form expression in (4.1). For a rigid sphere, however, no closed-form expression exists since the scattering term can be expressed only in the SHD. We therefore estimated the error by comparing the truncated Neumann Green's function $\hat{G}_{\text{N}}$ to a high-order Neumann Green's function. We can assume the error involved in using a high-order Neumann Green's function as a reference as opposed to the untruncated Neumann Green's function is small, due to the uniform convergence of the SHE [12]. In practice, we cannot choose very large values of L because of numerical difficulties involved in multiplying high order spherical Bessel and Hankel functions. For typical sphere radii and source-array distances, this allows us to reach L values of up to about 100 using SMIRgen, a MATLAB implementation of the SMIR method [14].

We evaluated the truncated (Neumann) Green's function at $K = 1024$ discrete values of k (denoted by $\dot{k}$), forming a set $\mathcal{K}$ corresponding to frequencies in the range

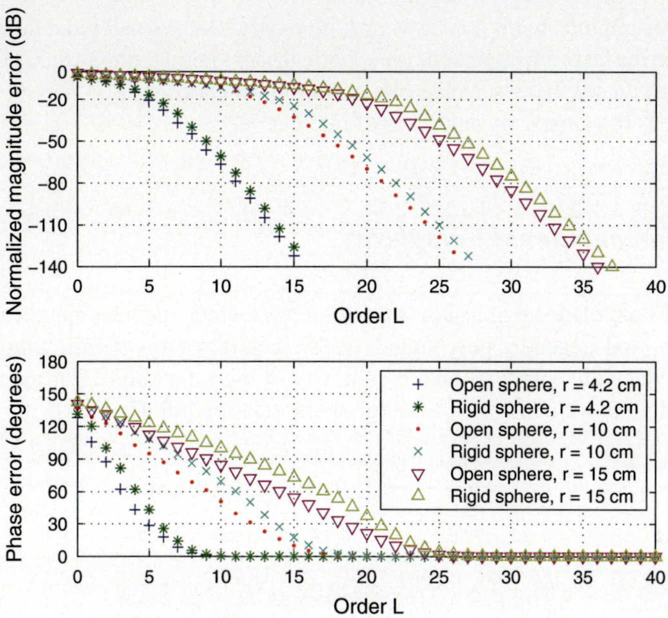

Fig. 4.5 Magnitude and phase errors involved in the truncation of the SHE in the Green's function (open sphere) and the Neumann Green's function (rigid sphere). The errors reduce rapidly beyond $L = k_{\max} r$, where here $k_{\max} = \frac{2\pi\, 8000}{c} \approx 147\ \text{m}^{-1}$

100 Hz–8 kHz,[5] and then calculated the normalized root-mean-square magnitude error ϵ_m and the root-mean-square phase error ϵ_p:

$$\epsilon_\text{m}(\mathbf{r}|\mathbf{r}_\text{s}, L) = \sqrt{\frac{1}{K} \sum_{\dot{k} \in \mathcal{K}} \frac{\left(\left| G_\text{N}(\mathbf{r}|\mathbf{r}_\text{s}, \dot{k}) \right| - \left| \hat{G}_\text{N}(\mathbf{r}|\mathbf{r}_\text{s}, \dot{k}, L) \right| \right)^2}{\left| G_\text{N}(\mathbf{r}|\mathbf{r}_\text{s}, \dot{k}) \right|^2}}, \tag{4.16}$$

$$\epsilon_\text{p}(\mathbf{r}|\mathbf{r}_\text{s}, L) = \sqrt{\frac{1}{K} \sum_{\dot{k} \in \mathcal{K}} \left(\angle G_\text{N}(\mathbf{r}|\mathbf{r}_\text{s}, \dot{k}) - \angle \hat{G}_\text{N}(\mathbf{r}|\mathbf{r}_\text{s}, \dot{k}, L) \right)^2}. \tag{4.17}$$

We averaged the magnitude and phase errors over 32 microphone positions uniformly distributed on the array and 50 random source positions at a fixed distance from the centre of the array.

The resulting average errors are given in Fig. 4.5, for both the open and rigid sphere cases. Three different sphere radii were used: $r = 4.2$ cm (the radius of the *Eigenmike* [24]), $r = 10$ cm and $r = 15$ cm. A source-array distance of 1 m was used; results for 1–5 m are omitted as they are essentially identical. It can be seen that

[5]Very low frequencies are omitted due to the fact that the spherical Hankel function $h_l(x)$ has a singularity around $x = 0$.

beyond a certain threshold, increases in L give only a very small reduction in error; this is due to the fast convergence of the spherical harmonic decomposition [12]. From Fig. 4.5, we can see that a sensible rule of thumb for choosing L is $L > \lceil 1.1\, k_{\max} r \rceil$ where $k_{\max}$ is the largest wavenumber of interest.

4.3.2 Computational Complexity

As the ATFs are made up of a sum over all orders l which includes spherical Hankel functions h_l and Legendre polynomials $\mathcal{P}_l$, we can make use of recursion relations over l to reduce the computational complexity of these functions. For the spherical Hankel function, we make use of the following relation [40, Eq. 6.69]

$$h_m^{(2)}(x) = \frac{2m-1}{x} h_{m-1}^{(2)}(x) - h_{m-2}^{(2)}(x), \; m \geq 2 \tag{4.18}$$

where

$$h_0^{(2)}(x) = -\frac{e^{-ix}}{ix}, \; h_1^{(2)}(x) = \frac{ie^{-ix}}{x^2} - \frac{e^{-ix}}{x}. \tag{4.19}$$

For the Legendre polynomial we use a similar recursion relation [1], known as Bonnet's recursion formula

$$\mathcal{P}_m(x) = \frac{2m-1}{m} x\mathcal{P}_{m-1}(x) - \frac{m-1}{m}\mathcal{P}_{m-2}(x), \; m \geq 2 \tag{4.20}$$

where $\mathcal{P}_0(x) = 1$ and $\mathcal{P}_1(x) = x$.

While replacing the exponential in (4.1) with a SHE does lead to an increase in computational complexity when computing the ATF for a single receiver (which is unavoidable in the rigid sphere case), this can have an advantage when simulating many receiver positions. For the conventional image method, we must compute the image positions and resulting response separately for each individual receiver. However, in the SMIR method the image positions are all computed with respect to the *centre* of our array, and therefore only once for all of the microphones in the array.

An alternative to (4.15) is obtained by changing the order of the summations in the ATF and computing the sum over all images only once, instead of once per receiver:

$$\begin{aligned} H(\mathbf{r}|\mathbf{r}_s, k) = &-ik \sum_{l=0}^{\infty} i^{-l} \sum_{m=-l}^{l} Y_{lm}(\Omega) \\ &\times \sum_{\mathbf{p}\in\mathcal{P}} \sum_{\mathbf{m}\in\mathcal{M}} \beta_{x_1}^{|m_x-p_x|} \beta_{x_2}^{|m_x|} \beta_{y_1}^{|m_y-p_y|} \beta_{y_2}^{|m_y|} \beta_{z_1}^{|m_z-p_z|} \beta_{z_2}^{|m_z|} \\ &\times b_l(k) h_l^{(2)}(k \left|\left|\mathbf{R}_{\mathbf{p},\mathbf{m}}\right|\right|) Y_{lm}^*(\angle \mathbf{R}_{\mathbf{p},\mathbf{m}}). \end{aligned} \tag{4.21}$$

The expression in (4.21) requires $O\left((N_\mathrm{i}+Q)(L+1)^2\right)$ operations per discrete frequency, where L is the maximum spherical harmonic order, N_i is the number of images and Q is the number of microphones, while the approach in (4.15) requires $O\left(N_\mathrm{i}Q(L+1)\right)$ operations per discrete frequency. Since the number of images N_i is typically very large, $(N_\mathrm{i}+Q)(L+1)^2 \approx N_\mathrm{i}(L+1)^2$. Assuming the operations in the two approaches are of similar complexity, it is therefore more efficient to use the expression in (4.15) for $Q < L+1$ and the expression in (4.21) for $Q > L+1$. Consequently the least computationally complex approach depends on the number of microphones Q and array radius r. In the remainder of this chapter we use the expression in (4.15); this is particularly appropriate in the applications in Sect. 4.4.2 where $Q = 2$ and in Sect. 4.4.3 where $Q = 1$.

4.3.3 Algorithm Summary

A summary of the SMIR method is presented in the form of pseudocode in Fig. 4.6. The variable $nsample$ denotes the number of samples in the AIR, N_o, the maximum reflection order, and fs, the sampling frequency.

The number of computations has been reduced by processing only half of the frequency spectrum because we know the AIR is real and the corresponding ATF is conjugate symmetric. The pseudocode necessary to compute the Hankel functions and Legendre polynomials is omitted here, since their computation is straightforward using recursion relations (4.18) and (4.20).

SMIRgen, a MATLAB/C++ implementation of the method in the form of a MEX-function, is available online [14].

4.4 Examples and Applications

In this section we give a number of examples that make use of the SMIR method. Wherever possible we compare the simulated results to theoretical results obtained using approximate models. These examples are given to illustrate and partially validate the SMIR method.

4.4.1 Diffuse Sound Field Energy

In statistical room acoustics (SRA), reverberant sound fields are modelled as diffuse sound fields, allowing for a statistical analysis of reverberation instead of computing each of the individual reflections. In this subsection, we compare a theoretical prediction of sound energy on the surface of a rigid sphere, based on a diffuse model of reverberation, to simulated results obtained using the SMIR method.

Fig. 4.6 Pseudocode for the SMIR method

```
P = {0,1}^3
M = {-N_m,···,0,···,N_m}^3
A = P × M

for (p,m) ∈ A do
  if |2m_x − p_x| + |2m_y − p_y| + |2m_z − p_z| ≤ N_o then
    R_{p,m} = [ x_s − 2p_x x_s − x_a + 2m_x L_x ;
                y_s − 2p_y y_s − y_a + 2m_y L_y ;
                z_s − 2p_z z_s − z_a + 2m_z L_z ]
    β(p,m) = β_{x1}^{|m_x−p_x|} β_{x2}^{|m_x|} β_{y1}^{|m_y−p_y|} β_{y2}^{|m_y|} β_{z1}^{|m_z−p_z|} β_{z2}^{|m_z|}
  else
    A = A \ {(p,m)}
  end if
end for

for k = 1 → nsample/2+1 do
  for l = 0 → L do
    if sphType = 'rigid' then
      Δ(k,l) = −i^{l+1} / (h_l^{(2)'}(kr)(kr)^2)
    else
      Δ(k,l) = i^l j_l(kr)
    end if
    Γ(k,l) = (−iki^{−l} / 4π) · Δ(k,l)
  end for
end for

for (p,m) ∈ A do
  if ||R_{p,m}|| + r < c·nsample/fs then
    for ang = 1 → Q do
      Θ = R̂_{p,m} · r̂(ang)
      for l = 0 → L do
        ϒ(ang,l) = P_l(Θ) · (2l+1)
      end for
    end for
    for k = 1 → nsample/2+1 do
      for l = 0 → L do
        Λ(k,l) = h_l(k||R_{p,m}||) · Γ(k,l)
      end for
    end for
    for ang = 1 → Q do
      for k = 1 → nsample/2+1 do
        for l = 0 → L do
          H(p,m,ang,k,l) = β(p,m) · ϒ(ang,l) · Λ(k,l)
        end for
        H(p,m,ang,k) = Σ_l H(p,m,ang,k,l)
      end for
    end for
  end if
end for
H(ang,k) = Σ_{(p,m)∈A} H(p,m,ang,k)
h(ang,n) = IFFT_R{H(ang,k)}
```

A diffuse sound field is composed of plane waves incident from all directions with equal probability and amplitude [20]. Using the scattering model previously introduced, we can determine the cross-correlation between the sound pressure at arbitrary positions $\mathbf{r}$ and $\mathbf{r}'$ on the surface of a sphere, due to a unit amplitude plane wave with a random uniformly distributed direction of arrival (see the Appendix for derivation) [15]:

$$C(\mathbf{r}, \mathbf{r}', k) = \sum_{l=0}^{\infty} |b_l(k)|^2 (2l+1) \mathcal{P}_l(\cos \Theta_{\mathbf{r},\mathbf{r}'}), \tag{4.22}$$

where $\Theta_{\mathbf{r},\mathbf{r}'}$ is the angle between $\mathbf{r}$ and $\mathbf{r}'$. In the open sphere case, it is shown in the Appendix that this simplifies to the well-known spatial domain expression [20, 31, 39] $\text{sinc}(k \left\|\mathbf{r} - \mathbf{r}'\right\|)$, where sinc denotes the unnormalized sinc function.

For the sound energy at a position $\mathbf{r}$ we substitute $\Theta_{\mathbf{r},\mathbf{r}'} = 0$ and find $C(\mathbf{r}, \mathbf{r}, k) = \sum_{l=0}^{\infty} |b_l(k)|^2 (2l+1)$. According to SRA theory [20, 39], for frequencies above the Schroeder frequency [20] the energy of the reverberant sound field H_r is then given by [39]

$$\begin{aligned}
\text{E}_\text{s}\left\{|H_\text{r}(\mathbf{r}, k)|^2\right\} &= \frac{1-\bar{\alpha}}{\pi A \bar{\alpha}} C(\mathbf{r}, \mathbf{r}, k) \\
&= \frac{1-\bar{\alpha}}{\pi A \bar{\alpha}} \sum_{l=0}^{\infty} |b_l(k)|^2 (2l+1),
\end{aligned} \tag{4.23}$$

where $\text{E}_\text{s}\left\{\cdot\right\}$ denotes spatial expectation, $\bar{\alpha}$ is the average wall absorption coefficient and A is the total wall surface area.

The above theoretical expression for the average reverberant energy can be compared to simulated results obtained using the SMIR method. We computed the spatial expectation using an average over 200 source-array positions, using the approach in Radlović et al. [31]: the array and source were kept in a fixed configuration (at a distance of 2 m from each other), which was then randomly rotated and translated. Both sources and microphones were kept at least half a wavelength from the boundaries of the room, helping to ensure the diffuseness of the reverberant sound field [20]. The reverberant component H_r of the ATFs was computed by subtracting the direct path H_d from the simulated ATFs.

The room dimensions were chosen as $6.4 \times 5 \times 4$ m, as in [31, 38], such that the ratio of the dimensions was $(1.6 : 1.25 : 1)$, as recommended in [18, 31] to approximate a diffuse sound field. The reverberation time T_{60} was set to 500 ms, giving an average wall absorption coefficient of $\bar{\alpha} = 0.2656$. We simulated AIRs with a length of 4096 samples at a sampling frequency of 8 kHz. We considered frequencies from 300 Hz to 4 kHz, well above the Schroeder frequency of $2000\sqrt{\frac{0.5}{6.4 \times 5 \times 4}} =$ 125 Hz, and the half-wavelength minimum distance is therefore 57 cm for a speed of sound of 343 m/s. We averaged the results over the 200 source-array positions and 32 microphone positions uniformly distributed on the array.

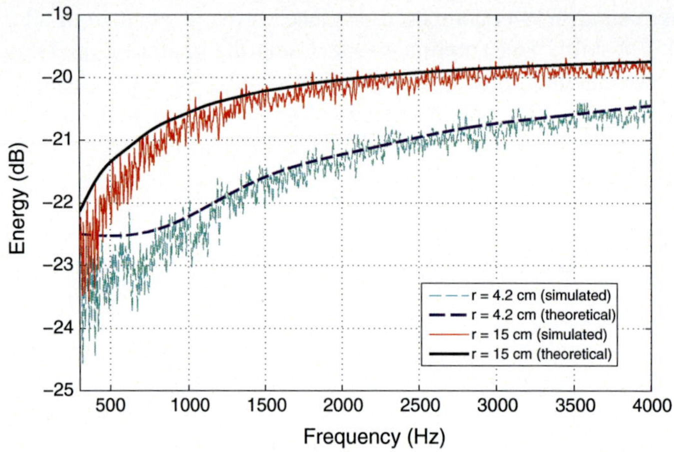

Fig. 4.7 Theoretical and simulated reverberant sound field energy on the surface of a rigid sphere, as a function of frequency for two array radii. The simulated results are averaged over 200 source-array positions, all at least half a wavelength from the room boundaries

In Fig. 4.7, we plot the theoretical and simulated energy of H_r as a function of frequency, for two array radii (4.2 and 15 cm). We note that, except at low frequencies, there is a good match between the theoretical diffuse field energy expression we derived and the results obtained using the SMIR method. At lower frequencies, the theoretical equation overestimates the energy; we hypothesize that this is due to the reverberant sound field not being fully diffuse.

4.4.2 Binaural Interaural Time and Level Differences

The topic of binaural sound and in particular head-related transfer functions (HRTFs) or head-related impulse responses (HRIRs) is of interest to researchers and engineers working on surround sound reproduction, who for example aim to reproduce spatial audio through a pair of stereo headphones. In addition, the psychoacoustic community is interested in the ability of the human brain to localize sound sources using only two ears.

Two binaural cues that contribute to sound source localization in humans are the interaural time difference (ITD) and the interaural level difference (ILD) [34]. The ITD measures the difference in arrival time of a sound at the two ears, and the ILD measures the difference in level of the sound at the two ears. In this example, we study the long-term cues assuming the source signal is spectrally white. Therefore, we can compute the cues directly using the simulated ATFs.

We used the SMIR method to simulate a simple HRTF by considering microphones placed at locations on a rigid sphere corresponding to ear positions on the human head. Although real HRTFs vary from individual to individual, depending on many factors including the head, torso and pinnae, many of the main characteristics of the HRTF are also exhibited by a simple rigid sphere ATF [9]. The representation of HRTFs using spherical harmonics was studied in [3, 10].

Whereas HRTFs do not normally include the effects of reverberation, and as a result typically sound artificial and provide poor cues for the perception of sound source distance [37], the SMIR method also allows for the inclusion of reverberation in HRIRs. In this case, they are then referred to as binaural room impulse responses (BRIRs). BRIRs are important for the analysis of the effects of reverberation on auditory perception, for example its impact on localization accuracy. Since rotational symmetry no longer necessarily holds once the room reflections are taken into account, the measurement of BRIRs must be done for every source-head position and orientation and is therefore very time-consuming. Simulating BRIRs allows us to more easily study the effects of early and late reflections on the binaural cues.

We begin by looking at ITDs in an anechoic environment, in order to illustrate the effect of the head in isolation. We compare simulated results to approximate theoretical results provided by a ray-tracing formula attributed to Woodworth and Schlosberg that looks at the distance travelled from the source to an observation point on the sphere, either in free-space if the observation point is on the near side of the sphere, or via a point of tangency if the observation point is on the far side [9].

The simulated results were obtained by using the SMIR method to generate HRIRs at a sampling frequency of 32 kHz, with a sphere radius of 8.75 cm and microphones placed at $(\theta, \phi) = (90°, 100°)$ (corresponding to the left ear) and $(\theta, \phi) = (90°, 260°)$ (corresponding to the right ear). The HRIRs were then band-pass filtered between 2.8 and 3.2 kHz.[6] The DOA was varied by rotating the source around the sphere at a fixed distance of 1 m and inclination of 90°. The simulated ITD was computed by determining the time delay that maximized the interaural cross-correlation between the two simulated and band-pass filtered HRIRs. The cross-correlation was interpolated using a second-order polynomial in order to obtain sub-sample delays.

In Fig. 4.8 we plot the ITDs as a function of direction of arrival, where 0° corresponds to the median plane on the front side of the sphere and 180° corresponds to the median plane on the back side of the sphere. As expected, as the DOA increases from 0° to 80° and the source gets closer to the ipsilateral ear, the ITD increases monotonically until it reaches its maximum at 80°, at which point the source is furthest from the contralateral ear. The ITD then decreases from 80° to 180° as the

---

[6]While the ray-tracing formula is frequency-independent, it has been shown [6] that ITDs actually exhibit some frequency dependence, and that because the ray-tracing concept applies to short wavelengths, this model yields only the high frequency time delay. Kuhn provides a more comprehensive discussion of this model and the frequency-dependence of ITDs [19]. It should be noted the simulation results in Fig. 4.8 are in broad agreement with Kuhn's measured results at 3.0 kHz.

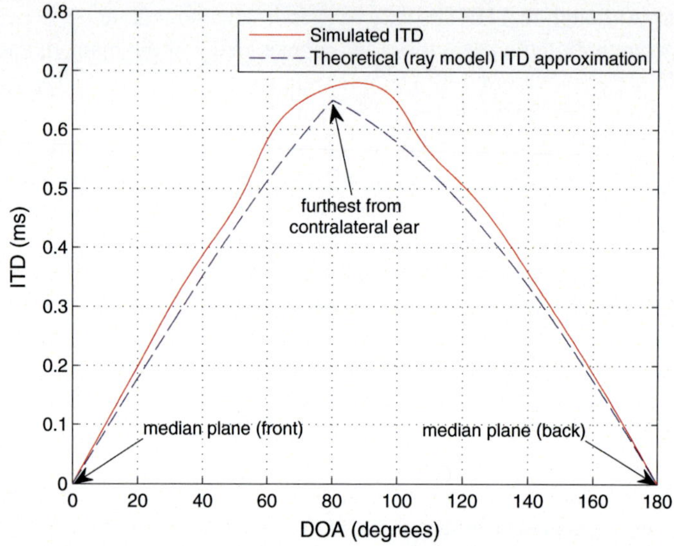

Fig. 4.8 Comparison of ITDs as a function of source DOA, in simulation and using the theoretical ray model approximation. The simulated ITDs are based on HRIRs computed using the SMIR method in an anechoic environment

source nears the median plane and gets closer to the contralateral ear. The response from 180° to 360° is not shown due to the symmetry about 180°. As we expect, the simulated results are reasonably close to the theoretical ray-tracing results [9], with a difference of less than 70 μs.

Using the SMIR method, we analyzed the ILDs in a reverberant environment under three scenarios: the sphere was either placed in the centre of the room with a DOA of 0° (where the source is equidistant from the two ears), or at a distance of approximately 0.5 m from one of the walls with DOAs of 0° and 100° (where the source is aligned with the left ear). In all three cases the source was placed at a distance of 1 m from the centre of the sphere. We chose a room size of $9 \times 5 \times 3$ m with a reverberation time T_{60} of 500 ms, and simulated BRIRs with a length of 4096 samples at a sampling frequency of 8 kHz.

In Figs. 4.9, 4.10 and 4.11 we plot the ILDs for the three above cases, as well as the ILDs we would obtain in an anechoic environment, which are entirely due to scattering. The ILDs were computed by taking the difference in magnitude between the left ear response and the right ear response. A negative ILD therefore indicates that the magnitude of the ipsilateral ear response is lower than that of the contralateral

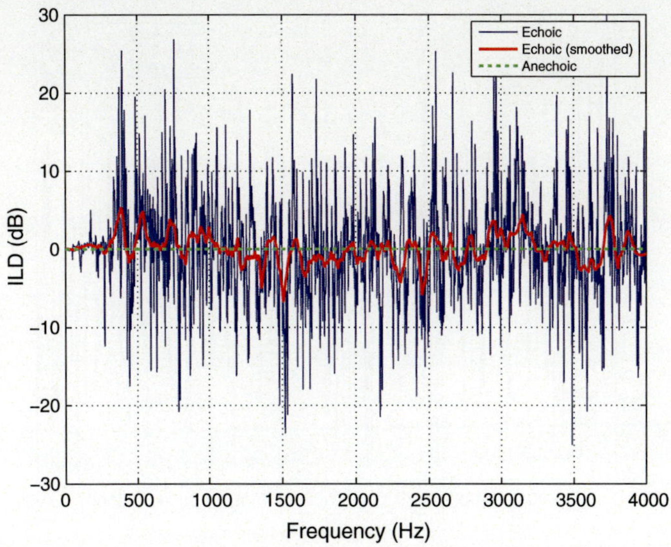

Fig. 4.9 Comparison of ILDs in echoic and anechoic environments, with the sphere placed in the centre of the room and a DOA of 0°. The ILDs are based on HRTFs (anechoic) and BRIRs (echoic) computed using the SMIR method

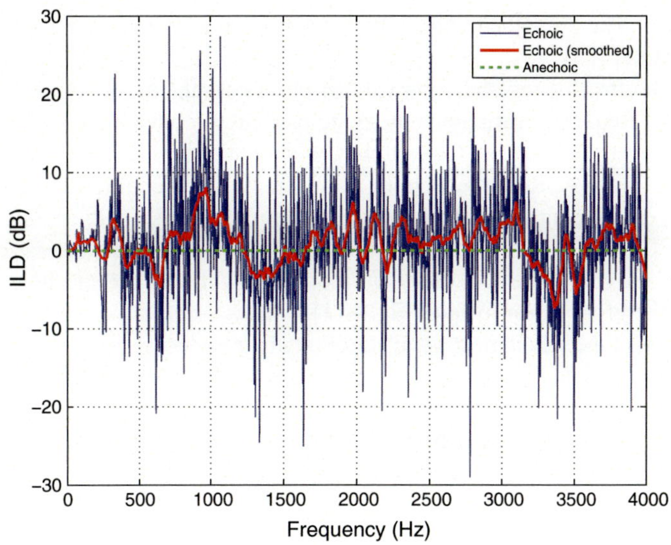

Fig. 4.10 Comparison of ILDs in echoic and anechoic environments, with the sphere placed near a room wall and a DOA of 0°

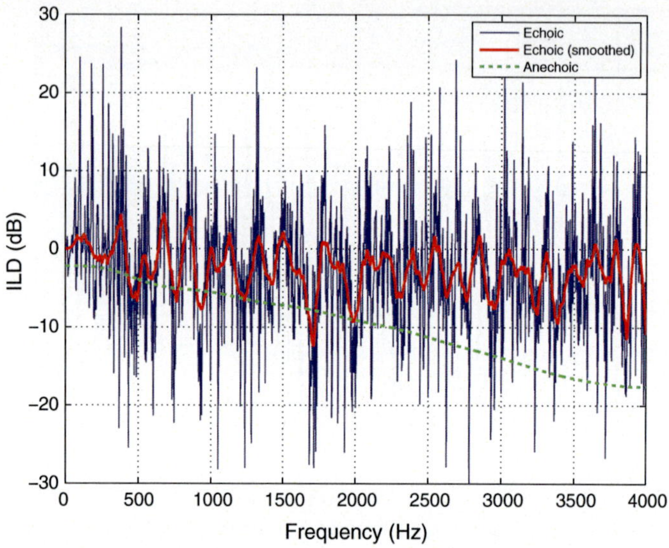

Fig. 4.11 Comparison of ILDs in echoic and anechoic environments, with the sphere placed near a room wall and a DOA of 100°

ear response. The smoothed echoic ILDs were obtained using a Savitzky-Golay smoothing filter [35].

The main effect of reverberation we can observe is the introduction of random frequency-to-frequency variations; these are particularly obvious when most of the reverberant energy is diffuse, for example, when the sphere is placed in the centre of the room (Fig. 4.9). Room reflections also increase the overall reverberant energy, particularly in the contralateral ear which receives less direct path energy, thus reducing the ILDs. This is especially noticeable when the contralateral ear is placed near a wall: the contralateral ear receives more energy than in the anechoic case and the ILD is therefore closer to zero (Fig. 4.11).

Placement of the sphere near a wall additionally introduces systematic distortions in the ILDs associated with the prominent early reflection from this wall. This is visible in Fig. 4.11 and most noticeably in Fig. 4.10.

All these effects have also been observed experimentally with a manikin by Shinn-Cunningham et al. [37]. The SMIR method is therefore an inexpensive way of predicting the effects of head movement and environmental changes (such as reverberation time) on HRTFs or BRIRs, without as much need for physical and acoustic measurements to be performed.

4.4.3 Mouth Simulator

The principle of reciprocity can often be advantageously used in room acoustics measurements. The principle states that ATFs are symmetric in the coordinates of the sound source and the observation point: *"If we put the sound source at* $\mathbf{r}$, *we observe at point* $\mathbf{r}_0$ *the same sound pressure as we did before at* $\mathbf{r}$, *when the sound source was at* $\mathbf{r}_0$" [20]. We can apply this principle to ATF simulations, and use the SMIR method to generate the ATF between one or more sources on a sphere and a single omnidirectional microphone placed away from the sphere.

A specific application of this is a mouth simulator: we model the head as a rigid sphere (as in Sect. 4.4.2) of radius r_h, and the mouth as an omnidirectional point source placed on this rigid sphere. This is straightforwardly implemented in the SMIR method by replacing the source position with the microphone position $\mathbf{r}_\mathrm{mic}$, the microphone position with the mouth position $\mathbf{r}_\mathrm{mouth} = (r_\mathrm{h}, \Omega_\mathrm{mouth})$, and the array position with the head position:

$$H(\mathbf{r}_\mathrm{mic}|\mathbf{r}_\mathrm{mouth}, k) = H(\mathbf{r} = \mathbf{r}_\mathrm{mouth}|\mathbf{r}_\mathrm{s} = \mathbf{r}_\mathrm{mic}, k).$$

As a result we can simulate the ATF between a mouth on a head, and a single microphone in free space. Repeated use of the algorithm allows for multiple receivers.

Although more accurate modelling of the head and mouth is possible using finite element or boundary element methods [5, 30] for example, the SMIR method is valuable for application to this problem due its comparative simplicity and the fact that, if desired, it can also take into account room reverberation. The SMIR method can, for example, be used as a mouth simulator in the evaluation of a speech enhancement algorithm [13], instead of the omnidirectional source model that is commonly used. While the diameter of the mouth plays an important role in determining the filter characteristic of the vocal tract [8], we assume for the purposes of the scattering model that the mouth is a point source.

As an illustration of this application, Fig. 4.12 shows the energy of the ATF between the mouth and a microphone as a function of microphone position at frequencies of 100 Hz and 3 kHz in an anechoic environment. The mouth was positioned on a sphere of radius 8.75 cm. Only two dimensions, x and y, are shown for brevity since the z dimension is identical to x and y. We observe that at 100 Hz there is no scattering and the radiation pattern is omnidirectional so that the sphere has little effect. At 3 kHz the effect of scattering starts to become more significant, and the energy at the back of the sphere is reduced while the energy at the front is increased. Finally the bright spot discussed in Sect. 4.2.3 is particularly apparent at the very back of the sphere in the bottom plot.

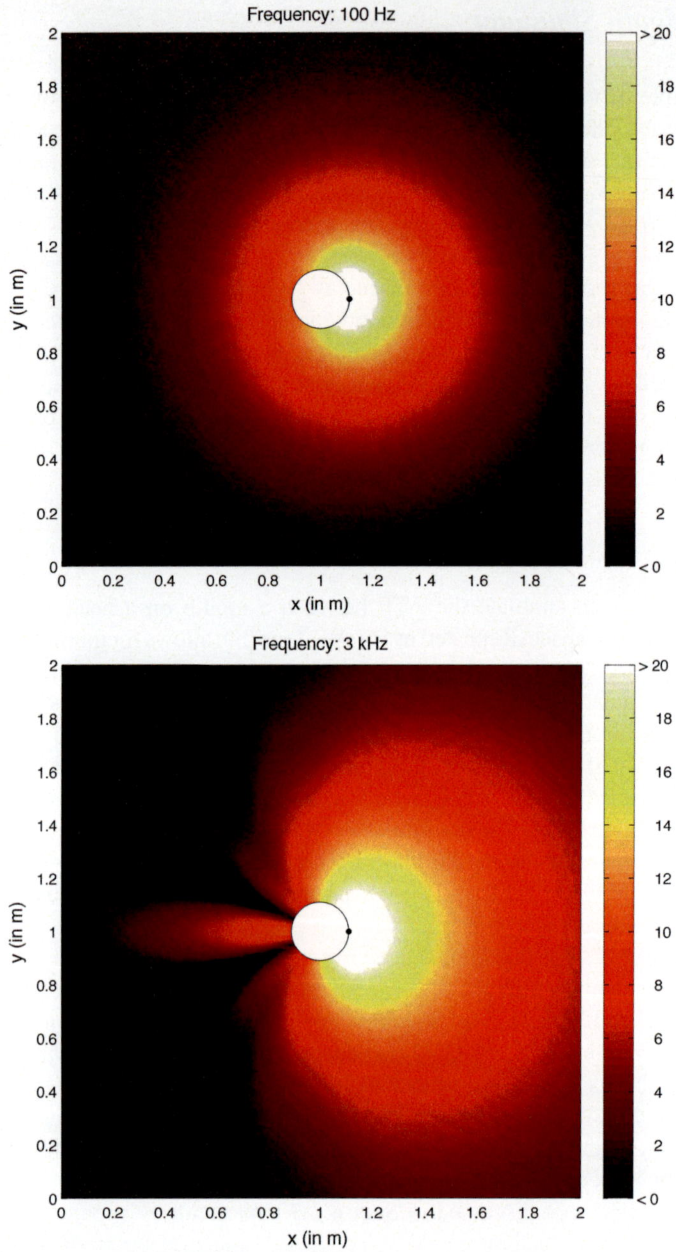

Fig. 4.12 Sound energy radiation pattern (in dB) at 100 Hz (*top*) and 3 kHz (*bottom*). The mouth position is denoted by a *black dot*

4.5 Chapter Summary and Conclusions

Spherical microphone arrays on a rigid baffle are of great interest, due to their numerical robustness and precisely calculable scattering effects. In order to analyze, work with and develop acoustic signal processing algorithms that make use of a spherical microphone array, a simulator is needed that can take into account the effects of the acoustic environment of the array as well as the scattering effects of the rigid spherical baffle. Accordingly, in this chapter the SMIR method was presented for the simulation of AIRs or ATFs for a rigid spherical microphone array in a reverberant environment.

We presented a scattering model used to model the rigid sphere, justifying its use with references to the literature, and provided an overview of the model's behaviour. We showed that the error with respect to the theoretical model can be controlled at the expense of increased computational complexity. Finally we provided a number of examples showing additional applications of this method.

Appendix: Spatial Correlation in a Diffuse Sound Field

The sound pressure at a position $\mathbf{r} = (r, \Omega)$ due to a unit amplitude plane wave incident from direction Ω_s is given by [40]

$$P(\mathbf{r}, \Omega_\mathrm{s}, k) = \sum_{l=0}^{\infty} \sum_{m=-l}^{l} 4\pi\varphi(\Omega_\mathrm{s}) b_l(k) Y_{lm}^*(\Omega_\mathrm{s}) Y_{lm}(\Omega), \tag{4.24}$$

where $\varphi(\Omega_\mathrm{s})$ is a random phase term and $|\varphi(\Omega_\mathrm{s})| = 1$. Assuming a diffuse sound field, the spatial cross-correlation between the sound pressure at two positions $\mathbf{r} = (r, \Omega)$ and $\mathbf{r}' = (r, \Omega')$ is given by:

$$\begin{aligned} C(\mathbf{r}, \mathbf{r}', k) &= \frac{1}{4\pi} \int_{\Omega_\mathrm{s} \in \mathcal{S}^2} P(\mathbf{r}, \Omega_\mathrm{s}, k) P^*(\mathbf{r}', \Omega_\mathrm{s}, k) d\Omega_\mathrm{s} \\ &= \frac{1}{4\pi} \int_{\Omega_\mathrm{s} \in \mathcal{S}^2} \sum_{l=0}^{\infty} \sum_{m=-l}^{l} 4\pi b_l(k) Y_{lm}^*(\Omega_\mathrm{s}) Y_{lm}(\Omega) \\ &\quad \times \sum_{l'=0}^{\infty} \sum_{m'=-l'}^{l'} 4\pi b_{l'}^*(kr) Y_{l'm'}(\Omega_\mathrm{s}) Y_{l'm'}^*(\Omega') d\Omega_\mathrm{s}. \end{aligned}$$

Using the orthonormality property of the spherical harmonics in (2.18) and the addition theorem in (2.23), we eliminate the cross terms followed by the sum over m and obtain

$$
\begin{aligned}
C(\mathbf{r}, \mathbf{r}', k) &= \frac{1}{4\pi} \sum_{l=0}^{\infty} \sum_{m=-l}^{l} (4\pi)^2 |b_l(k)|^2 Y_{lm}(\Omega) Y_{lm}^*(\Omega') \\
&= \frac{1}{4\pi} \sum_{l=0}^{\infty} (4\pi)^2 |b_l(k)|^2 \frac{2l+1}{4\pi} \mathcal{P}_l(\cos \Theta_{\mathbf{r},\mathbf{r}'}) \\
&= \sum_{l=0}^{\infty} |b_l(k)|^2 (2l+1) \mathcal{P}_l(\cos \Theta_{\mathbf{r},\mathbf{r}'}),
\end{aligned} \tag{4.25}
$$

where $\Theta_{\mathbf{r},\mathbf{r}'}$ is the angle between $\mathbf{r}$ and $\mathbf{r}'$.

In the open sphere case, we can derive a simplified expression for $C(\mathbf{r}, \mathbf{r}', k)$. Firstly, we note that the expression in (4.25) is real, and therefore, for a reason which will soon become clear, $C(\mathbf{r}, \mathbf{r}', k)$ can advantageously be expressed as

$$
C(\mathbf{r}, \mathbf{r}', k) = -\Im \left\{ -i \sum_{l=0}^{\infty} |b_l(k)|^2 (2l+1) \mathcal{P}_l(\cos \Theta_{\mathbf{r},\mathbf{r}'}) \right\}, \tag{4.26}
$$

where $\Im$ denotes the imaginary part of a complex number. By substituting the open sphere mode strength $b_l(k) = i^l j_l(kr)$ into (4.26), we obtain

$$
C(\mathbf{r}, \mathbf{r}', k) = -\Im \left\{ -i \sum_{l=0}^{\infty} j_l^2(kr) (2l+1) \mathcal{P}_l(\cos \Theta_{\mathbf{r},\mathbf{r}'}) \right\}. \tag{4.27}
$$

Using $\Re\{h_l^{(2)}(kr)\} = j_l(kr)$, where $\Re$ denotes the real part of a complex number, we can now write (4.27) as

$$
\begin{aligned}
C(\mathbf{r}, \mathbf{r}', k) &= -\Im \left\{ -i \sum_{l=0}^{\infty} j_l(kr) \left[h_l^{(2)}(kr) - i\Im\{h_l^{(2)}(kr)\} \right] (2l+1) \mathcal{P}_l(\cos \Theta_{\mathbf{r},\mathbf{r}'}) \right\} \\
&= -\Im \left\{ -i \sum_{l=0}^{\infty} j_l(kr) h_l^{(2)}(kr) (2l+1) \mathcal{P}_l(\cos \Theta_{\mathbf{r},\mathbf{r}'}) \right\} \\
&\quad + \Im \left\{ \underbrace{\sum_{l=0}^{\infty} j_l(kr) \Im\{h_l^{(2)}(kr)\} (2l+1) \mathcal{P}_l(\cos \Theta_{\mathbf{r},\mathbf{r}'})}_{\star} \right\}.
\end{aligned} \tag{4.28}
$$

As the expression marked with a $\star$ is real, its imaginary part is zero and (4.28) can be simplified to

$$
C(\mathbf{r}, \mathbf{r}', k) = -\Im \left\{ -i \sum_{l=0}^{\infty} j_l(kr) h_l^{(2)}(kr) (2l+1) \mathcal{P}_l(\cos \Theta_{\mathbf{r},\mathbf{r}'}) \right\}. \tag{4.29}
$$

Finally, using (4.7) and (4.8), we obtain the well-known spatial domain result for two omnidirectional receivers in a diffuse sound field [20, 31, 39]:

$$\begin{aligned} C(\mathbf{r}, \mathbf{r}', k) &= -\Im\left\{\frac{e^{-ik||\mathbf{r}-\mathbf{r}'||}}{k\,||\mathbf{r}-\mathbf{r}'||}\right\} \\ &= \frac{\sin(k\,||\mathbf{r}-\mathbf{r}'||)}{k\,||\mathbf{r}-\mathbf{r}'||}. \end{aligned} \tag{4.30}$$

References

1. Abramowitz, M., Stegun, I.A. (eds.): Handbook of Mathematical Functions with Formulas, Graphs, and Mathematical Tables. Dover Publications, New York (1972)
2. Allen, J.B., Berkley, D.A.: Image method for efficiently simulating small-room acoustics. J. Acoust. Soc. Am. **65**(4), 943–950 (1979)
3. Avni, A., Rafaely, B.: Sound localization in a sound field represented by spherical harmonics. In: Proceedings 2nd International Symposium on Ambisonics and Spherical Acoustics, pp. 1–5. Paris, France (2010)
4. Betlehem, T., Poletti, M.A.: Sound field reproduction around a scatterer in reverberation. In: Proceedings of IEEE International Conference on Acoustics, Speech and Signal Processing (ICASSP), pp. 89–92 (2009). doi:10.1109/ICASSP.2009.4959527
5. Botteldoore, D.: Finite-difference time-domain simulation of low-frequency room acoustic problems. J. Acoust. Soc. Am. **98**(6), 3302–3308 (1995)
6. Brown, C., Duda, R.: A structural model for binaural sound synthesis. IEEE Trans. Speech Audio Process. **6**(5), 476–488 (1998). doi:10.1109/89.709673
7. Chen, Z., Maher, R.C.: Addressing the discrepancy between measured and modeled impulse responses for small rooms. In: Proceedings of Audio Engineering Society Convention (2007)
8. Deller, J.R., Proakis, J.G., Hansen, J.H.L.: Discrete-Time Processing of Speech Signals. MacMillan, New York (1993)
9. Duda, R.O., Martens, W.L.: Range dependence of the response of a spherical head model. J. Acoust. Soc. Am. **104**(5), 3048–3058 (1998). doi:10.1121/1.423886
10. Evans, M.J., Angus, J.A.S., Tew, A.I.: Analyzing head-related transfer function measurements using surface spherical harmonics. J. Acoust. Soc. Am. **104**(4), 2400–2411 (1998)
11. Gumerov, N., Duraiswami, R.: Modeling the effect of a nearby boundary on the HRTF. In: Proceedings of IEEE International Conference on Acoustics, Speech and Signal Processing (ICASSP), vol. 5, pp. 3337–3340 (2001). doi:10.1109/ICASSP.2001.940373
12. Gumerov, N.A., Duraiswami, R.: Fast Multipole Methods for the Helmholtz Equation in Three Dimensions. Elsevier, Oxford (2005)
13. Habets, E.A.P., Benesty, J.: A two-stage beamforming approach for noise reduction and dereverberation. IEEE Trans. Audio, Speech, Lang. Process. **21**(5), 945–958 (2013)
14. Jarrett, D.P.: Spherical microphone array impulse response (SMIR) generator. http://www.ee.ic.ac.uk/sap/smirgen/
15. Jarrett, D.P., Habets, E.A.P., Thomas, M.R.P., Gaubitch, N.D., Naylor, P.A.: Dereverberation performance of rigid and open spherical microphone arrays: theory and simulation. In: Proceedings Joint Workshop on Hands-Free Speech Communication and Microphone Arrays (HSCMA), pp. 145–150. Edinburgh (2011)
16. Jarrett, D.P., Habets, E.A.P., Thomas, M.R.P., Naylor, P.A.: Simulating room impulse responses for spherical microphone arrays. In: Proc. IEEE Intl. Conf. on Acoustics, Speech and Signal Processing (ICASSP), pp. 129–132. Prague, Czech Republic (2011)

17. Jarrett, D.P., Habets, E.A.P., Thomas, M.R.P., Naylor, P.A.: Rigid sphere room impulse response simulation: algorithm and applications. J. Acoust. Soc. Am. **132**(3), 1462–1472 (2012)
18. Knudsen, V., Harris, C.: Acoustical Designing in Architecture. Wiley, New York (1950)
19. Kuhn, G.F.: Model for the interaural time differences in the azimuthal plane. J. Acoust. Soc. Am. **62**(1), 157–167 (1977). doi:10.1121/1.381498
20. Kuttruff, H.: Room Acoustics, 4th edn. Taylor and Francis, London (2000)
21. Lehmann, E., Johansson, A.: Diffuse reverberation model for efficient image-source simulation of room impulse responses. IEEE Trans. Audio Speech Lang. Process. **18**(6), 1429–1439 (2010). doi:10.1109/TASL.2009.2035038
22. Li, Z., Duraiswami, R.: Hemispherical microphone arrays for sound capture and beamforming. In: Proceedings of IEEE Workshop on Applications of Signal Processing to Audio and Acoustics, pp. 106–109 (2005)
23. Logan, N.A.: Survey of some early studies of the scattering of plane waves by a sphere. Proc. IEEE **53**(8), 773–785 (1965). doi:10.1109/PROC.1965.4055
24. Meyer, J., Agnello, T.: Spherical microphone array for spatial sound recording. In: Proc. Audio Engineering Society Convention, pp. 1–9. New York (2003)
25. Meyer, J., Elko, G.: A highly scalable spherical microphone array based on an orthonormal decomposition of the soundfield. In: Proceedings of IEEE International Conference on Acoustics, Speech and Signal Processing (ICASSP), vol. 2, pp. 1781–1784 (2002)
26. Meyer, J., Elko, G.W.: Position independent close-talking microphone. Signal Process. **86**(6), 1254–1259 (2006). doi:10.1016/j.sigpro.2005.05.036
27. mh acoustics LLC: The Eigenmike microphone array. http://www.mhacoustics.com/mh_acoustics/Eigenmike_microphone_array.html
28. Morse, P.M., Ingard, K.U.: Theoretical Acoustics. International Series in Pure and Applied Physics. McGraw Hill, New York (1968)
29. Peterson, P.M.: Simulating the response of multiple microphones to a single acoustic source in a reverberant room. J. Acoust. Soc. Am. **80**(5), 1527–1529 (1986)
30. Pietrzyk, A.: Computer modeling of the sound field in small rooms. In: Proceedings of AES International Conference on Audio, Acoustics and Small Spaces, vol. 2, pp. 24–31. Copenhagen (1998)
31. Radlović, B.D., Williamson, R., Kennedy, R.: Equalization in an acoustic reverberant environment: robustness results. IEEE Trans. Speech Audio Process. **8**(3), 311–319 (2000)
32. Rafaely, B.: Plane-wave decomposition of the pressure on a sphere by spherical convolution. J. Acoust. Soc. Am. **116**(4), 2149–2157 (2004)
33. Rafaely, B.: Analysis and design of spherical microphone arrays. IEEE Trans. Speech Audio Process. **13**(1), 135–143 (2005). doi:10.1109/TSA.2004.839244
34. Sandel, T.T., Teas, D.C., Feddersen, W.E., Jeffress, L.A.: Localization of sound from single and paired sources. J. Acoust. Soc. Am. **27**(5), 842–852 (1955). doi:10.1121/1.1908052
35. Savitzky, A., Golay, M.J.E.: Smoothing and differentiation of data by simplified least squares procedures. Anal. Chem. **36**(8), 1627–1639 (1964). doi:10.1021/ac60214a047
36. Sengupta, D.L.: The sphere. In: J.J. Bowman, T.B.A. Senior, P.L.E. Uslenghi (eds.) Electromagnetic and Acoustic Scattering by Simple Shapes, pp. 353–415. North-Holland (1969) (chap. 10)
37. Shinn-Cunningham, B.G., Kopco, N., Martin, T.J.: Localizing nearby sound sources in a classroom: binaural room impulse responses. J. Acoust. Soc. Am. **117**(5), 3100–3115 (2005). doi:10.1121/1.1872572
38. Talantzis, F., Ward, D.B.: Robustness of multichannel equalization in an acoustic reverberant environment. J. Acoust. Soc. Am. **114**(2), 833–841 (2003)
39. Ward, D.B.: On the performance of acoustic crosstalk cancellation in a reverberant environment. J. Acoust. Soc. Am. **110**, 1195–1198 (2001)
40. Williams, E.G.: Fourier Acoustics: Sound Radiation and Nearfield Acoustical Holography, 1st edn. Academic Press, London (1999)

Chapter 5
Acoustic Parameter Estimation

Acoustic parameter estimation, including the estimation of quantities that describe the sound field, is a major field of research within acoustic signal processing. Considerable research interest has focused on the estimation of parameters relating to sound sources, such as the number of active sound sources and their direction of arrival DOA. It can also be of use to estimate room acoustic parameters, such as reverberation time, or parameters which relate to both the acoustic environment and the sound source, such as the signal-to-diffuse energy ratio.

The estimated acoustic parameters can potentially provide additional a priori information to speech enhancement algorithms, thereby improving their performance. In this chapter, we present methods for estimating two such parameters using spherical microphone arrays: the DOA of one or more sources (Sect. 5.1) and the signal-to-diffuse ratio (SDR) (Sect. 5.2).

5.1 Direction of Arrival Estimation

In this section, we address the problem of two-dimensional DOA estimation with spherical microphone arrays. It should be noted that we refer to this problem as *two-dimensional* to refer to estimation of azimuth and elevation but not range, so that the source-array distance does not figure in our estimation. In this sense, DOA estimation differs from source localization in which the source position is localized in all three dimensions. As a consequence, we emphasize that referring to DOA as a two-dimensional problem does not imply that the source position is confined to a two-dimensional space. DOA estimates are most commonly used in conjunction with a beamformer (see Chaps. 6 and 7), in order to determine the steering direction.

Portions of Sect. 5.1.5 and the Appendix were first published in [13], and are reproduced here with the author's permission.

D. Jarrett et al., *Theory and Applications of Spherical Microphone Array Processing*, Springer Topics in Signal Processing 9,
DOI 10.1007/978-3-319-42211-4_5

They can also be used to identify noise sources (for example, in a vehicle or aircraft engine), or for automated camera steering.

One-dimensional DOA estimation (azimuth-only or elevation-only) with microphone arrays has been extensively studied, often based on techniques originally developed for antenna arrays. Notable approaches include time difference of arrival (TDOA)–based methods such as GCC-PHAT [19], subspace-based methods such as ESPRIT [33] or MUSIC [34], and steered response power (SRP) methods [10].

In this section, we will present a number of DOA estimation methods in the spherical harmonic domain (SHD): an SRP approach, the pseudointensity vector method, and two subspace-based approaches. We will also present a comparative performance evaluation of the SRP and pseudointensity vector methods in noisy reverberant environments. It should be noted that TDOA–based methods are normally unsuitable for spherical microphone arrays with small radii, due to the insufficient spacing between individual microphones.

Other SHD DOA estimation methods have recently been proposed, but detailed discussion of these is beyond the scope of this chapter. In [23, 25], the authors proposed DOA estimation methods suitable for highly reverberant environments based on a direct-path dominance test. In [15], a method for moving source tracking was proposed. In [8], a method of DOA estimation in the presence of multiple sources using a clustering approach was proposed. A method of DOA estimation in the presence of two simultaneous plane waves was also proposed based on a B-format microphone signal [44].

5.1.1 Problem Formulation

We consider the sound pressure $P(k, \mathbf{r}_q)$ captured by an array of Q microphones at positions $\mathbf{r}_q$, $q \in \{1, \dots, Q\}$, where k denotes the wavenumber.[1] We assume that I plane waves impinge upon the array with DOAs $\Omega_\iota = (\theta_\iota, \phi_\iota)$ (where the angle θ_ι represents inclination, and the angle ϕ_ι represents azimuth), $\iota \in \{1, \dots, I\}$; these DOAs are the quantities we wish to estimate. The sound pressure $P(k, \mathbf{r}_q)$ at a position $\mathbf{r}_q$ can then be expressed as

$$P(k, \mathbf{r}_q) = \sum_{\iota=1}^{I} X(k, \mathbf{r}_q, \Omega_\iota) S_\iota(k) + V(k, \mathbf{r}_q), \tag{5.1}$$

where $X(k, \mathbf{r}_q, \Omega_\iota)$ denotes the sound pressure at a position $\mathbf{r}_q$ due to a *unit amplitude* plane wave originating from a direction Ω_ι, $S_\iota(k)$ denotes the amplitude of the ιth plane wave, and $V(k, \mathbf{r}_q)$ denotes a noise signal, used to model sensor noise and (optionally) reverberation.

[1] The dependency on time is omitted for brevity. In practice, the signals acquired using a spherical microphone array are usually processed in the short-time Fourier transform domain, as explained in Sect. 3.1, where the discrete frequency index is denoted by ν.

When using spherical microphone arrays, it is convenient to work in the SHD [22, 30], instead of the spatial domain. We assume error-free spatial sampling, and refer the reader to Chap. 3 for information on spatial sampling and aliasing. By applying the complex spherical harmonic transform (SHT) to the signal model in (5.1), we obtain the SHD signal model

$$P_{lm}(k) = \sum_{\iota=1}^{I} X_{lm}(k, \Omega_\iota) S_\iota(k) + V_{lm}(k), \tag{5.2}$$

where $P_{lm}(k)$, $X_{lm}(k, \Omega_\iota)$ and $V_{lm}(k)$ are respectively the SHTs of the spatial domain signals $P(k, \mathbf{r}_q)$, $X(k, \mathbf{r}_q, \Omega_\iota)$ and $V(k, \mathbf{r}_q)$, as defined in (3.6), and are referred to as *eigenbeams* to reflect the fact that the spherical harmonics are eigensolutions of the wave equation in spherical coordinates [39]. The order and degree of the spherical harmonics are respectively denoted as l and m.

The eigenbeams $P_{lm}(k)$, $X_{lm}(k, \Omega_\iota)$ and $V_{lm}(k)$ are a function of the frequency-dependent mode strength $b_l(k)$. The mode strength captures the dependence of the eigenbeams on the array properties (radius, microphone type, configuration). Mode strength expressions for two common types of arrays, the open and rigid arrays with omnidirectional microphones, are given in Sect. 3.4.2. To cancel this dependence, we divide the eigenbeams by the mode strength (as in [27]), thus giving mode strength compensated eigenbeams, and the SHD signal model is then written as

$$\widetilde{P}_{lm}(k) = \left[\sqrt{4\pi} b_l(k)\right]^{-1} P_{lm}(k) \tag{5.3a}$$

$$= \sum_{\iota=1}^{I} \widetilde{X}_{lm}(k, \Omega_\iota) S_\iota(k) + \widetilde{V}_{lm}(k), \tag{5.3b}$$

where $\widetilde{P}_{lm}(k)$, $\widetilde{X}_{lm}(k, \Omega_\iota)$ and $\widetilde{V}_{lm}(k)$ respectively denote the eigenbeams $P_{lm}(k)$, $X_{lm}(k, \Omega_\iota)$ and $V_{lm}(k)$ after mode strength compensation. It should be noted that the mode strength compensation can also be applied directly to the microphone signals. The design of such equalization filters is beyond the scope of this book. The eigenbeam $\widetilde{X}_{lm}(k, \Omega_\iota)$, which is due to a unit amplitude plane wave, is given by dividing the expression in (3.22a) by the mode strength $b_l(k)$ to yield

$$\widetilde{X}_{lm}(\Omega_\iota) = Y_{lm}^*(\Omega_\iota), \tag{5.4}$$

where $Y_{lm}(\Omega_\iota)$ denotes the complex spherical harmonic[2] of order l and degree m evaluated at an angle Ω_ι, as defined in (2.14). We observe that $\widetilde{X}_{lm}(\Omega_\iota)$ is independent of frequency due to the mode strength compensation process, and depends only on the DOA Ω_ι.

---

[2]If the real SHT is applied instead of the complex SHT, the complex spherical harmonics Y_{lm} used throughout this chapter should be replaced with the real spherical harmonics R_{lm}, as defined in Sect. 3.3.

For convenience, we express (5.3b) in vector/matrix notation, where the SHD vectors have length $N = (L+1)^2$, the total number of eigenbeams from order $l = 0$ to $l = L$:

$$\widetilde{\boldsymbol{p}}(k) = \widetilde{\mathbf{X}}(\boldsymbol{\Omega})\,\mathbf{s}(k) + \widetilde{\boldsymbol{v}}(k), \tag{5.5}$$

where

$$\widetilde{\boldsymbol{p}}(k) = \left[\widetilde{P}_{00}(k),\ \widetilde{P}_{1(-1)}(k),\ \widetilde{P}_{10}(k),\ \widetilde{P}_{11}(k),\ \widetilde{P}_{2(-2)}(k), \ldots, \widetilde{P}_{LL}(k)\right]^{\mathrm{T}}, \tag{5.6}$$

$$\widetilde{\boldsymbol{v}}(k) = \left[\widetilde{V}_{00}(k),\ \widetilde{V}_{1(-1)}(k),\ \widetilde{V}_{10}(k),\ \widetilde{V}_{11}(k),\ \widetilde{V}_{2(-2)}(k), \ldots, \widetilde{V}_{LL}(k)\right]^{\mathrm{T}}, \tag{5.7}$$

$$\mathbf{s}(k) = \left[S_1(k),\ S_2(k),\ \ldots,\ S_I(k)\right]^{\mathrm{T}}, \tag{5.8}$$

$$\boldsymbol{\Omega} = \left[\Omega_1,\ \Omega_2,\ \ldots,\ \Omega_I\right]^{\mathrm{T}}, \tag{5.9}$$

$\widetilde{\mathbf{X}}(\boldsymbol{\Omega})$ is the array manifold matrix given by

$$\widetilde{\mathbf{X}}(\boldsymbol{\Omega}) = \left[\widetilde{\boldsymbol{x}}(\Omega_1)\,\middle|\,\widetilde{\boldsymbol{x}}(\Omega_2)\,\middle|\,\cdots\,\middle|\,\widetilde{\boldsymbol{x}}(\Omega_I)\right] \tag{5.10}$$

and $\widetilde{\boldsymbol{x}}(\Omega_\iota)$ is the array manifold vector given by

$$\widetilde{\boldsymbol{x}}(\Omega_\iota) = \left[\widetilde{X}_{00}(\Omega_\iota),\ \widetilde{X}_{1(-1)}(\Omega_\iota),\ \ldots,\ \widetilde{X}_{LL}(\Omega_\iota)\right]^{\mathrm{T}}, \iota \in \{1, \ldots, I\}. \tag{5.11}$$

The ordering of the orders and degrees in (5.11) is consistent with the ordering commonly used in Ambisonics. The Ambisonic channel number (ACN) corresponding to order l and degree m is given by $l(l+1)+m$ [4].

5.1.2 Steered Response Power

A conventional DOA estimation method in the spatial domain involves computing a map of the SRP, which is obtained by determining the output power of a beamformer as a function of the steering direction. Assuming that only a single source is active in each time-frequency bin (W-disjoint orthogonality [3, 29]), the DOA is given by the direction with maximum power.

The output of an Lth-order SHD beamformer steered in a direction Ω_{u} can be expressed as [32, Eq. 11.41]

$$Z(k, \Omega_{\mathrm{u}}) = \sum_{l=0}^{L} \sum_{m=-l}^{l} W_{lm}^{*}(k, \Omega_{\mathrm{u}}) P_{lm}(k) \tag{5.12}$$

where $W_{lm}(k, \Omega_{\mathrm{u}})$ denotes the beamformer weights. The SRP method can be implemented using any beamformer, such as the signal-dependent minimum variance distortionless response (MVDR) beamformer used in [37], or the signal-independent plane-wave decomposition (PWD) beamformer used in [14]. The reader is referred to Chaps. 6 and 7 for details of beamformers that could be suitable for this purpose.

A narrowband power map $\mathcal{P}_{\mathrm{SRP}}(k, \Omega)$ at a wavenumber k is given by

$$\mathcal{P}_{\mathrm{SRP}}(k, \Omega) = |Z(k, \Omega)|^2 . \tag{5.13}$$

If only a single source is active at a particular time instant, a smoothed broadband power map $\mathcal{P}_{\mathrm{SRP}}(\Omega)$ can be obtained by averaging the power $|Z(k, \Omega)|^2$ over wavenumbers[3] from K_{s} to K_{e}:

$$\mathcal{P}_{\mathrm{SRP}}(\Omega) = \int_{K_{\mathrm{s}}}^{K_{\mathrm{e}}} \beta_Z(k)\, |Z(k, \Omega)|^2 \mathrm{d}k, \tag{5.14}$$

where $\beta_Z(k)$ is a frequency-dependent weighting function, such as an A-weighting function [9]. The weighting function should usually be chosen based on the spectral characteristics of the active source and noise. The power map can additionally be smoothed over time; if the source is moving, this smoothing trades off DOA estimation accuracy against time resolution.

A narrowband estimate $\hat{\Omega}_{\mathrm{s}}(k)$ of the DOA is then given by the direction with highest power:

$$\hat{\Omega}_{\mathrm{s}}(k) = \arg\max_{\Omega} \mathcal{P}_{\mathrm{SRP}}(k, \Omega). \tag{5.15}$$

Alternatively, a broadband estimate $\hat{\Omega}_{\mathrm{s}}$ is obtained from the smoothed broadband power map as

$$\hat{\Omega}_{\mathrm{s}} = \arg\max_{\Omega} \mathcal{P}_{\mathrm{SRP}}(\Omega). \tag{5.16}$$

When multiple plane waves are present ($I > 1$), the DOAs are given by the I maxima of the power map $\mathcal{P}_{\mathrm{SRP}}$.

5.1.3 Intensity-Based Method

In acoustics, sound intensity is a measure of the flow of sound energy through a unit area perpendicular to the direction of sound propagation per unit time [20]. The (active) intensity vector $\boldsymbol{\mathcal{I}}(\mathbf{r}, k)$ at a position $\mathbf{r}$, introduced in Sect. 2.4, indicates the magnitude and direction of the transport of acoustical energy, and is related to the (scalar) sound pressure P and the particle velocity vector $\mathbf{v}$ via (2.33) [6]

[3] In practice, since the processing is performed in the short-time Fourier transform domain, this integral is approximated with a sum over discrete frequency indices ν.

$$\mathcal{I}(\mathbf{r}, k) = \frac{1}{2}\Re\left\{P(\mathbf{r}, k) \cdot \mathbf{v}^*(\mathbf{r}, k)\right\}, \tag{5.17}$$

where $\Re\{\cdot\}$ denotes the real part of a complex number.

We again assume that at each time-frequency instant, only a single plane wave is present with DOA Ω_1. The particle velocity vector is then given by (2.34) [5, p. 31]

$$\mathbf{v}(\mathbf{r}, k) = -\frac{P(\mathbf{r}, k)}{\rho_0 c}\mathbf{u}(\mathbf{r}, k), \tag{5.18}$$

where c is the speed of sound in the medium, ρ_0 is the ambient density, and $\mathbf{u}$ is a unit vector pointing from $\mathbf{r}$ towards the acoustic source (in a direction Ω_1). As a result, the intensity vector points in the opposite direction to the vector $\mathbf{u}$.

Various techniques for measuring the intensity vector are discussed in Sect. 2.4. Estimating the intensity vector using a spherical microphone array composed of pressure microphones is of practical interest, since the numerous pressure microphones can then also be used for acoustic signal enhancement.

In [14], it was proposed to estimate the DOA based on the direction of the *pseudointensity vector*, which is conceptually similar to the intensity vector, but is calculated using the zero- and first-order eigenbeams $P_{lm}(k)$ $(l = 0, 1)$. The pseudointensity vector $\mathbf{I}(k)$ is defined as

$$\mathbf{I}(k) = \frac{1}{2}\Re\left\{\widetilde{P}_{00}(k)\begin{bmatrix} P_x^*(k) \\ P_y^*(k) \\ P_z^*(k) \end{bmatrix}\right\}, \tag{5.19}$$

where the first term, $\widetilde{P}_{00}(k)$, is equal (as shown in the Appendix) to the sound pressure that would be measured were a microphone to be placed at the centre of the array, and the second term is a scaled estimate of the particle velocity vector. The components $P_x(k)$, $P_y(k)$ and $P_z(k)$ of this vector are dipole signals steered such that the dipoles are aligned with the x, y and z axes. These dipole signals approximate the pressure gradient, which is proportional to the particle velocity vector for a single plane wave [21, 46].

The dipole signals $P_x(k)$, $P_y(k)$ and $P_z(k)$ are obtained by forming a linear combination of the rotated first-order eigenbeams:

$$P_a(k) = \sum_{m=-1}^{1} \frac{Y_{1m}(\Omega_a)}{b_1(k)} P_{1m}(k) \tag{5.20a}$$

$$= \sum_{m=-1}^{1} Y_{1m}(\Omega_a)\widetilde{P}_{1m}(k), \qquad a \in \{x, y, z\}, \tag{5.20b}$$

where the division by $b_1(k)$ is required to make the beam patterns independent of wavenumber, and the steering angles Ω_a required to align the dipoles with the x, y and z axes are given by

$$\Omega_x = (\theta_x, \phi_x) = (\pi/2, \pi), \tag{5.21a}$$

$$\Omega_y = (\theta_y, \phi_y) = (\pi/2, -\pi/2), \tag{5.21b}$$

$$\Omega_z = (\theta_z, \phi_z) = (\pi, 0), \tag{5.21c}$$

in the spherical coordinate system introduced in Sect. 2.2.

A narrowband estimate $\hat{\mathbf{u}}(k)$ of the unit vector pointing from the centre of the array to the active source is then computed as

$$\hat{\mathbf{u}}(k) = -\frac{\mathbf{I}(k)}{||\mathbf{I}(k)||}, \tag{5.22}$$

where $||\cdot||$ denotes the 2-norm (Euclidian norm).

These estimates are suitable for applications where DOA estimates are required for every time-frequency instant; for example, in beamforming (Sect. 9.1). If only a single source is active at a particular time instant, a broadband estimate with reduced variance is obtained using a pseudointensity vector $\mathbf{I}$ averaged across wavenumbers from K_s to K_e:

$$\mathbf{I} = \int_{K_\mathrm{s}}^{K_\mathrm{e}} \beta_I(k)\mathbf{I}(k)\mathrm{d}k, \tag{5.23}$$

where $\beta_I(k)$ is a weighting function similar to $\beta_Z(k)$ in (5.14). Note that even with $\beta_I(k) = 1, \forall k$, a higher weight is implicitly given to the pseudointensity vectors with the highest norm, which are considered to be more reliable for DOA estimation.

The pseudointensity vectors can also be averaged over time, where the time averaging interval will depend on whether multiple sources are present and whether they are static or moving. A related method was proposed in [15] for single source tracking using spherical microphone arrays, based on an adaptive principal component analysis of particle velocity vector estimates. In addition, a pseudointensity vector–based method for DOA estimation with multiple sources was proposed in [8].

The performance of the pseudointensity vector method presented above may deteriorate in the presence of strong coherent reflections. A pseudointensity vector–based method that is more robust to such reflections was proposed in [23].

The pseudointensity vector method is related to previous intensity vector–based DOA estimation approaches [2, 26]. However, the intensity vector was computed using an acoustic vector sensor, or using the Ambisonic B-format signals in the field of Directional Audio Coding (DirAC) [2]. The B-format signals are often measured directly (using an omnidirectional microphone and three dipole microphones) or with a three or four omnidirectional microphone grid. The eigenbeams used here for DOA estimation are computed using all of the microphones in a spherical array, of which there are typically a few dozen, thus providing more robustness to noise that is incoherent in the SHD (either spatially incoherent noise, or diffuse noise).

5.1.4 Subspace Methods

In subspace-based DOA estimation methods, the vector space of the covariance matrix of the noisy eigenbeams $\widetilde{\mathbf{p}}(k)$ is decomposed into two orthogonal subspaces: the signal subspace, and the noise subspace. The $N \times N$ covariance matrix $\boldsymbol{\Phi}_{\widetilde{\mathbf{p}}}(k)$ of the noisy eigenbeams $\widetilde{\mathbf{p}}(k)$ is given by

$$\boldsymbol{\Phi}_{\widetilde{\mathbf{p}}}(k) = \mathrm{E}\left\{\widetilde{\mathbf{p}}(k)\widetilde{\mathbf{p}}^{\mathrm{H}}(k)\right\} \tag{5.24a}$$

$$= \widetilde{\mathbf{X}}(\boldsymbol{\Omega})\boldsymbol{\Phi}_{\mathbf{s}}(k)\widetilde{\mathbf{X}}^{\mathrm{H}}(\boldsymbol{\Omega}) + \boldsymbol{\Phi}_{\widetilde{\mathbf{v}}}(k), \tag{5.24b}$$

where $\mathrm{E}\{\cdot\}$ denotes mathematical expectation, $\boldsymbol{\Phi}_{\mathbf{s}}(k) = \mathrm{E}\left\{\mathbf{s}(k)\mathbf{s}^{\mathrm{H}}(k)\right\}$ is the $I \times I$ covariance matrix of the plane wave amplitudes $\mathbf{s}(k)$ and $\boldsymbol{\Phi}_{\widetilde{\mathbf{v}}}(k) = \mathrm{E}\left\{\widetilde{\mathbf{v}}(k)\widetilde{\mathbf{v}}^{\mathrm{H}}(k)\right\}$ is the $N \times N$ noise covariance matrix. The noise covariance matrix is assumed to be diagonal, or equivalently, the noise is assumed to be spatially incoherent or spatially diffuse. In practice, the expectation in (5.24) is computed using a temporal averaging process.

We assume that the I plane waves are *not* mutually coherent; they may, however, be partially coherent. As a result, the covariance matrix $\boldsymbol{\Phi}_{\mathbf{s}}(k)$ has full rank, and the covariance matrix $\boldsymbol{\Phi}_{\widetilde{\mathbf{p}}}(k)$ can be decomposed in terms of its N eigenvalues $\lambda_{\boldsymbol{\Phi}_{\widetilde{\mathbf{p}}},\jmath}(k)$ and eigenvectors $\boldsymbol{\xi}_{\jmath}(k)$, $\jmath \in \{1, \ldots, N\}$:

$$\boldsymbol{\Phi}_{\widetilde{\mathbf{p}}}(k) = \sum_{\jmath=1}^{N} \lambda_{\boldsymbol{\Phi}_{\widetilde{\mathbf{p}}},\jmath}(k)\boldsymbol{\xi}_{\jmath}(k)\boldsymbol{\xi}_{\jmath}^{\mathrm{H}}(k). \tag{5.25}$$

The eigenvalues are assumed to be arranged in decreasing order. The eigenvectors can then be separated into two orthogonal subspaces. The first I eigenvectors, corresponding to the I largest eigenvalues, define a *signal subspace*

$$\mathbf{U}_s(k) = \left[\boldsymbol{\xi}_1(k) \,\middle|\, \boldsymbol{\xi}_2(k) \,\middle|\, \cdots \,\middle|\, \boldsymbol{\xi}_I(k)\right], \tag{5.26}$$

while the remaining eigenvectors, corresponding to the $N - I$ smallest eigenvalues, define a noise subspace

$$\mathbf{U}_\nu(k) = \left[\boldsymbol{\xi}_{I+1}(k) \,\middle|\, \boldsymbol{\xi}_{I+2}(k) \,\middle|\, \cdots \,\middle|\, \boldsymbol{\xi}_N(k)\right]. \tag{5.27}$$

The signal subspace contains both a signal component and a noise component, whereas the noise subspace does not contain any signal component [45]. We assume the number of plane waves I is known a priori; this quantity can be estimated using the algorithm presented in [41], for example.

Subspace-based DOA estimation methods take advantage of the fact that the eigenvectors $\boldsymbol{\xi}_\jmath$, $\jmath \in \{1, \ldots, I\}$ are linear combinations of the array manifold vectors $\widetilde{\mathbf{x}}(\Omega_\iota)$, $\iota \in \{1, \ldots, I\}$ [39], that is,

$$\text{span}\left\{\mathbf{U}_s(k)\right\} = \text{span}\left\{\widetilde{\mathbf{X}}(\boldsymbol{\Omega})\right\}. \tag{5.28}$$

On the other hand, any vector in the signal subspace is orthogonal to the noise subspace.

If two or more of the I plane waves are mutually coherent, as they may be in a reverberant environment where some of the plane waves correspond to reflections of the directional signal(s), the covariance matrix $\boldsymbol{\Phi}_\mathbf{s}(k)$ is rank deficient, that is,

$$\text{rank}\left\{\boldsymbol{\Phi}_\mathbf{s}(k)\right\} < I. \tag{5.29}$$

In this case, the array manifold vectors $\widetilde{\mathbf{x}}(\Omega_\iota)$ no longer span the signal subspace, and only the DOAs corresponding to the plane waves that are not mutually coherent can be estimated. Spatial and frequency smoothing techniques to overcome this problem with subspace-based methods are proposed in [17, 18, 37].

EB-MUSIC

The multiple signal classification (MUSIC) technique [35] is based on the fact that the noise subspace is orthogonal to the array manifold vectors. Hence, the projection of the vectors $\widetilde{\mathbf{x}}(\Omega_\iota)$, $\iota \in \{1, \ldots, I\}$ onto the noise subspace is zero. In the SHD, the EB-MUSIC pseudospectrum is defined as [32]

$$\mathcal{P}_{\text{MUSIC}}(k, \Omega) = \frac{1}{\widetilde{\mathbf{x}}^{\text{H}}(\Omega)\mathbf{U}_\nu(k)\mathbf{U}_\nu^{\text{H}}(k)\widetilde{\mathbf{x}}(\Omega)} \tag{5.30}$$

The Ω values corresponding to the I maxima of $\mathcal{P}_{\text{MUSIC}}(k, \Omega)$ are the DOAs we wish to estimate. A sample plot of the EB-MUSIC pseudospectrum obtained using eigenbeams up to order $L = 3$ is shown in Fig. 5.1 for plane waves at $(135°, 90°)$ and $(45°, 270°)$.

This approach provides accurate DOA estimates, but requires an exhaustive search of the two-dimensional solution space, making it computationally inefficient [32].

EB-ESPRIT

The estimation of signal parameters via rotational invariance techniques (ESPRIT) was first proposed in the spatial domain in [33]. A similar technique, EB-ESPRIT (eigenbeam-ESPRIT), was later proposed in the SHD in [40, 41].

In the spatial domain, ESPRIT takes advantage of a shift invariance property; in a similar way, EB-ESPRIT is based on a recurrence relation for the associated Legendre functions $\mathcal{P}_{lm}$:

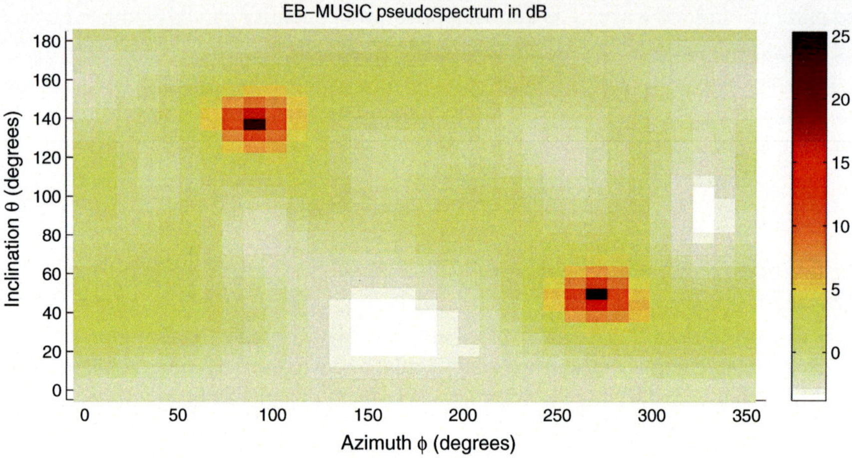

Fig. 5.1 Sample plot of the EB-MUSIC pseudospectrum as a function of inclination and azimuth for two plane waves. The two peaks correspond to the DOAs of the plane waves

$$2m \cot(\theta)\mathcal{P}_{lm}(\cos\theta) = (m-l-1)(l+m)\mathcal{P}_{l(m-1)}(\cos\theta) - \mathcal{P}_{l(m+1)}(\cos\theta). \quad (5.31)$$

In this section, we work only with eigenbeams of a single, fixed order L ($L \geq 2$), of which there are $2L+1$. Accordingly, quantities that are based on these eigenbeams are denoted with a subscript L. Let $\widetilde{\boldsymbol{x}}_L(\Omega)$ denote the last $2L+1$ elements of the array manifold vector $\widetilde{\boldsymbol{x}}(\Omega)$, or equivalently,

$$\widetilde{\boldsymbol{x}}_L(\Omega) = \left[\widetilde{X}_{L(-L)}(\Omega), \ldots, \widetilde{X}_{L(L-1)}(\Omega), \widetilde{X}_{LL}(\Omega)\right]^{\mathrm{T}}. \quad (5.32)$$

We can now form three overlapping subarrays of length $2L-1$, as illustrated in Fig. 5.2, with manifold vectors given by

$$\widetilde{\boldsymbol{x}}_L^{(i_l)}(\Omega) = \boldsymbol{\Delta}^{(i_l)}\, \mathbf{D}_0\, \widetilde{\boldsymbol{x}}_L(\Omega),\; i_l \in \{-1, 0, 1\}, \quad (5.33)$$

where the selection matrices $\boldsymbol{\Delta}^{(-1)}$, $\boldsymbol{\Delta}^{(0)}$ and $\boldsymbol{\Delta}^{(1)}$ extract the first, middle and last $2L-1$ elements from $\mathbf{D}_0\, \widetilde{\boldsymbol{x}}_L(\Omega)$, and

$$\mathbf{D}_0 = \operatorname{diag}\left\{(-1)^L, (-1)^{L-1}, \ldots, (-1)^0, 1, \ldots, 1^{L-1}, 1^L\right\}. \quad (5.34)$$

The $(2L-1) \times I$ subarray manifold matrices are given by

$$\widetilde{\mathbf{X}}_L^{(i_l)}(\boldsymbol{\Omega}) = \left[\widetilde{\boldsymbol{x}}_L^{(i_l)}(\Omega_1)\middle| \cdots \middle|\widetilde{\boldsymbol{x}}_L^{(i_l)}(\Omega_I)\right],\; i_l \in \{-1, 0, 1\}. \quad (5.35)$$

Using the recurrence relation for the associated Legendre functions (5.31), it can be shown that the subarray manifold matrices are related via the recurrence relation [41]

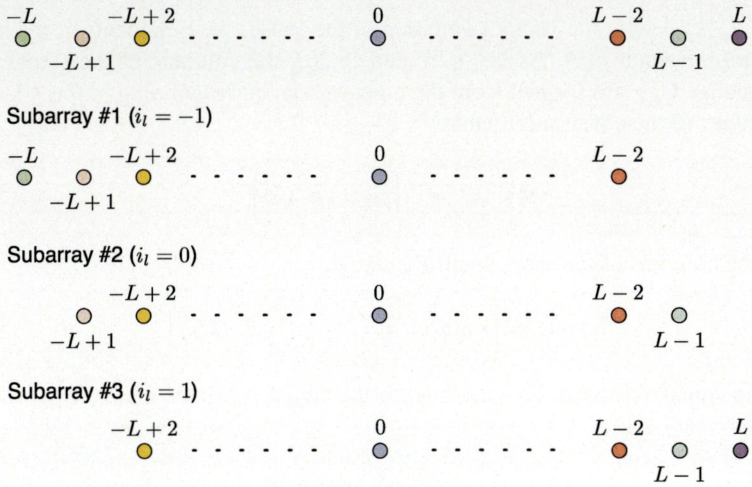

Fig. 5.2 Subarrays for EB-ESPRIT algorithm. The eigenbeams are all of order L; the numbers shown indicate the degree of the eigenbeams

$$\mathbf{D}_1 \widetilde{\mathbf{X}}_L^{(0)}(\boldsymbol{\Omega}) = \mathbf{D}_2 \widetilde{\mathbf{X}}_L^{(-1)}(\boldsymbol{\Omega}) \boldsymbol{\Phi}(\boldsymbol{\Omega}) + \mathbf{D}_3 \widetilde{\mathbf{X}}_L^{(1)}(\boldsymbol{\Omega}) \boldsymbol{\Phi}^*(\boldsymbol{\Omega}), \tag{5.36}$$

where the $I \times I$ steering matrix $\boldsymbol{\Phi}(\boldsymbol{\Omega})$ is given by

$$\boldsymbol{\Phi}(\boldsymbol{\Omega}) = \text{diag}\left\{\mu(\Omega_1), \ldots, \mu(\Omega_I)\right\}, \tag{5.37}$$

and

$$\mu(\Omega_\iota) = \tan\theta_\iota \, e^{-i\phi_\iota}, \iota \in \{1, \ldots, I\}. \tag{5.38}$$

The $(2L-1) \times (2L-1)$ matrices $\mathbf{D}_1$, $\mathbf{D}_2$ and $\mathbf{D}_3$ are defined as [41]

$$\mathbf{D}_1 = 2\,\text{diag}\left\{\frac{|m|}{a_{Lm}}\right\}, \tag{5.39a}$$

$$\mathbf{D}_2 = \text{diag}\left\{\frac{(m-L-1)(L+m)}{a_{L(m-1)}}\right\}, \tag{5.39b}$$

$$\mathbf{D}_3 = \text{diag}\left\{\frac{1}{a_{L(-L+2)}}, \frac{1}{a_{L(-L+1)}}, \ldots, \frac{1}{a_{L0}}, \frac{-1}{a_{L1}}, \frac{1}{a_{L2}}, \ldots, \frac{1}{a_{LL}}\right\}, \tag{5.39c}$$

with $m \in \{-(L-1), \ldots, L-1\}$ and where a_{Lm} is the angular-independent portion of the spherical harmonics, i.e.,

$$a_{Lm} = \sqrt{\frac{2L+1}{4\pi} \frac{(L-m)!}{(L+m)!}}. \tag{5.40}$$

Let $\widetilde{\boldsymbol{p}}_L(k)$ denote the vector composed of the last $2L+1$ elements of the noisy eigenbeams vector $\widetilde{\boldsymbol{p}}(k)$. As in (5.25) and (5.26), the columns of the signal subspace matrix $\mathbf{U}_{s,L}$ are formed from the eigenvectors corresponding to the I largest eigenvalues of the covariance matrix

$$\boldsymbol{\Phi}_{\widetilde{\boldsymbol{p}}_L}(k) = \mathrm{E}\left\{\widetilde{\boldsymbol{p}}_L(k)\widetilde{\boldsymbol{p}}_L^{\mathrm{H}}(k)\right\}. \tag{5.41}$$

Since the columns of the array manifold matrix

$$\widetilde{\mathbf{X}}_L(\boldsymbol{\Omega}) = \left[\widetilde{\boldsymbol{x}}_L(\Omega_1)\left|\widetilde{\boldsymbol{x}}_L(\Omega_2)\right|\cdots\left|\widetilde{\boldsymbol{x}}_L(\Omega_I)\right.\right] \tag{5.42}$$

span the signal subspace, $\widetilde{\mathbf{X}}_L$ is related to the signal subspace matrix $\mathbf{U}_{s,L}$ via the expression

$$\mathbf{U}_{s,L}(k) = \widetilde{\mathbf{X}}_L(\boldsymbol{\Omega})\mathbf{T}(k), \tag{5.43}$$

where $\mathbf{T}$ may be any non-singular $I \times I$ matrix. The subarray signal subspace matrices $\mathbf{U}_{s,L}^{(i_l)}$ are then computed as

$$\begin{aligned}\mathbf{U}_{s,L}^{(i_l)}(k) &= \boldsymbol{\Delta}^{(i_l)}\,\mathbf{U}_{s,L}(k), \qquad i_l \in \{-1,0,1\} &\text{(5.44a)}\\ &= \boldsymbol{\Delta}^{(i_l)}\,\widetilde{\mathbf{X}}_L(\boldsymbol{\Omega})\mathbf{T}(k). &\text{(5.44b)}\end{aligned}$$

As a result, the array manifold matrix recurrence relation (5.36) can be expressed as

$$\mathbf{D}_1\mathbf{U}_{s,L}^{(0)}(k) = \underbrace{\left[\mathbf{D}_2\mathbf{U}_{s,L}^{(-1)}(k)\middle|\,\mathbf{D}_3\mathbf{U}_{s,L}^{(1)}(k)\right]}_{\star}\begin{bmatrix}\boldsymbol{\Psi}^{\mathrm{T}}(k)\\ \boldsymbol{\Psi}^{\mathrm{H}}(k)\end{bmatrix}, \tag{5.45}$$

where

$$\boldsymbol{\Psi}(k) = \mathbf{T}^{-1}\boldsymbol{\Phi}(k)\mathbf{T} \tag{5.46}$$

and the star ($\star$) identifies a matrix for later cross-referencing.

Finally, by solving (5.45) in a least squares or total least squares sense, an estimate of $\boldsymbol{\Psi}$ can be obtained. As the eigenvalues of $\boldsymbol{\Psi}$, denoted as $\lambda_{\boldsymbol{\Psi},\iota}$, are the elements of $\boldsymbol{\Phi}$, an estimate of the azimuths ϕ_ι and inclinations θ_ι of the I plane waves is obtained as

$$\begin{aligned}\theta_\iota(k) &= \tan^{-1}\left[\left|\lambda_{\boldsymbol{\Psi},\iota}(k)\right|\right], \qquad \iota \in \{1,\ldots,I\} &\text{(5.47a)}\\ \phi_\iota(k) &= \arg\left[\lambda_{\boldsymbol{\Psi},\iota}(k)\right]. &\text{(5.47b)}\end{aligned}$$

From (5.47a) it is clear that a source with an inclination of $\theta = \pi/2$ cannot be localized. In [38], the authors propose to electronically rotate the eigenbeams using Wigner-D functions, thereby avoiding this issue, providing that some a priori knowledge of the source inclination is available. EB-ESPRIT also suffers from a sign ambiguity problem, since a particular eigenvalue $\lambda_{\Psi,\iota}$ can correspond to a source at $(\theta_\iota, \phi_\iota)$ or $(\pi - \theta_\iota, \pi + \phi_\iota)$. This ambiguity can be resolved by computing the SRP in both these directions, and choosing the direction with maximum power, as proposed in [38], resulting in only a minimal increase in computational complexity.

EB-ESPRIT is able to resolve a maximum of $L-1$ plane waves, that is, we require $L \geq I+1$, due to the dimensions of the starred matrix in (5.45). However, if instead of considering only the eigenbeams of a single order L (of which there are $2L+1$), we consider all eigenbeams of order $l \in \{1, \ldots, L\}$, of which there are $(L+1)^2 - 1$, the dimensions of this matrix can be significantly increased. As shown in [38], the number of resolvable plane waves then increases from $L-1$ to $\lfloor L^2/2 \rfloor$, where $\lfloor \cdot \rfloor$ denotes the floor operator.

EB-ESPRIT can also be applied to eigenbeams that have not been mode strength compensated, that is, $P_{lm}(k)$ instead of $\widetilde{P}_{lm}(k)$. This is due to the fact that, providing EB-ESPRIT is applied to eigenbeams of a *single* order L (as we have done in this section), the mode strength terms $b_L(k)$ cancel in the recurrence relation (5.36).

5.1.5 *Results*

It is interesting now to evaluate the performance of the algorithms presented in Sects. 5.1.2 and 5.1.3, namely the SRP and pseudointensity vector methods. Their performance can be quantified by calculating the angle ϵ between a unit vector pointing in the correct direction $\mathbf{u}$, and a unit vector $\hat{\mathbf{u}}$ pointing in the direction estimated by either of the two methods. The angular estimation error ϵ is then given by

$$\epsilon = \cos^{-1}(\mathbf{u}^{\mathrm{T}}\hat{\mathbf{u}}). \tag{5.48}$$

Experimental Setup

The performance of the algorithms was evaluated in a simulated environment, where the true source positions were known precisely. Simulated impulse responses were obtained using SMIRgen [12], an acoustic impulse response (AIR) simulator for spherical microphone arrays based on the algorithm presented in Chap. 4.

For these simulations, a $Q = 32$ microphone array with radius 4.2 cm was placed near the centre of an acoustic space with dimensions $10 \times 8 \times 12$ m in which a single source was present. The source signal consisted of a white Gaussian noise sequence. Spatio-temporally white Gaussian noise was added to the individual microphone signals in order to obtain an input signal-to-incoherent-noise ratio (iSINR) of 20 dB at the microphone closest to the source, that is, the microphone with the highest iSINR. The signals were processed in the short-time Fourier transform (STFT) domain with a sampling frequency of 8 kHz and a frame length of 64 ms with a 50 % overlap between successive frames.

The SRP method was implemented using a PWD beamformer, as presented in Sect. 6.3.1.1. The power map and pseudointensity vectors were averaged over 5 time frames, i.e., 192 ms of data. No weighting was applied in (5.14) and (5.23), that is, $\beta_Z(k) = \beta_I(k) = 1, \forall k$, and the average was computed over all frequencies up to 4 kHz. The same number of eigenbeams were used for the SRP as for the pseudointensity vector method, such that the limit $L = 1$.

Discussion

Two simulations are now discussed. In the first simulation, the reverberation time T_{60} was varied between 0 (anechoic room) and 600 ms while the source-array distance was fixed at 2.5 m. The room boundary reflection coefficients were computed from the desired reverberation times using Sabin-Franklin's formula [28]. With such a configuration, reverberation times between 300 and 600 ms corresponded to direct-to-reverberant energy ratios between approximately 10 and 0 dB. In the second simulation the source-array distance ranged between 1 and 3 m while the reverberation time was fixed at 450 ms.

A statistical analysis of the results of these simulations is shown in Fig. 5.3, based on Monte Carlo simulations with 100 runs. For each run a new DOA was randomly selected from a uniform angular distribution around the sphere. The accuracy of the pseudointensity vector method can be seen to be significantly higher than that of the SRP method with a small number of beams (266). For a larger number of beams (3962), the pseudointensity vector method still outperforms the SRP method, but by a smaller margin. This is still the case even as the source-array distance increases above 2 m and the reverberation time increases above 450 ms.

As expected, the accuracy of the pseudointensity vector method increases as the source-array distance and reverberation time decrease, since both these changes lead to an increase in the direct-to-reverberant energy ratio. In a purely diffuse sound field, the average intensity vector is zero [11]. When the direct-to-reverberant energy ratio is high, the reverberant field is mostly diffuse and causes little bias in the DOA estimates once they have been averaged over frequency (and optionally over time).

The pseudointensity method requires the computation of the four zero- and first-order eigenbeams $P_{00}(k)$, $P_{1(-1)(k)}$, $P_{10}(k)$ and $P_{11}(k)$, as well as three weighted averages $Z_x(k)$, $Z_y(k)$ and $Z_z(k)$ of these eigenbeams. The SRP method, on the other hand, requires computation of these eigenbeams (and potentially more eigenbeams if $L \geq 2$), and additionally as many weighted averages of these eigenbeams as required to yield the desired power map resolution.

From a computational complexity point of view, the pseudointensity vector method is therefore equivalent to the SRP method with three beams. However, as we have seen, very many more beams are necessary for the SRP method to obtain reasonable accuracy. We note that in practice, it is not efficient to steer hundreds or thousands of beams indiscriminately in all directions: a coarse grid approach can be taken at first, to determine the DOA within $\pm 30°$, for example, and then a finer grid can be applied to the area of interest, thus reducing the amount of unnecessary detail in directions where the acoustic source cannot be located (based on the results of the first search).

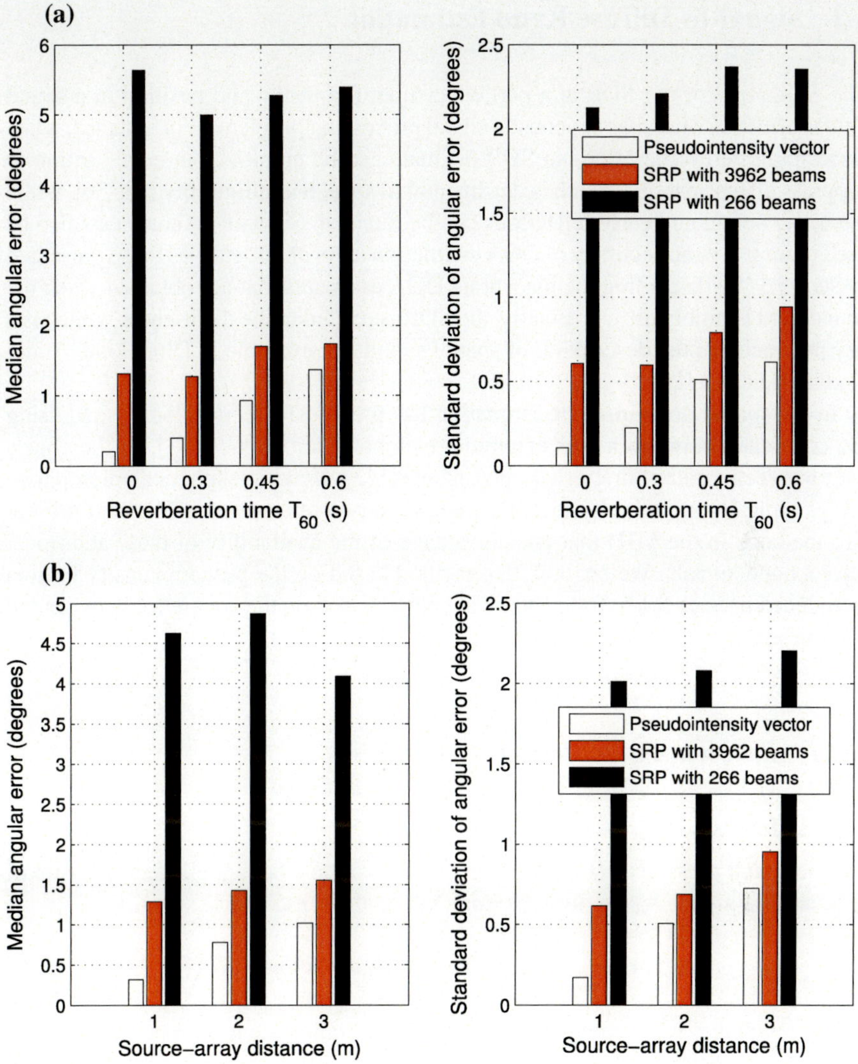

Fig. 5.3 Median and standard deviation of the angular errors for the SRP and pseudointensity vector methods, as a function of reverberation time (**a**) and source-array distance (**b**). In **a** the source-array distance is 2.5 m and in **b** the reverberation time is 450 ms; both of these conditions ensure that the direct-to-reverberant energy ratio remains above 0 dB. Copyright © Daniel Jarrett. Used with permission

5.2 Signal-to-Diffuse Ratio Estimation

The estimation of the SDR at a particular time, frequency and position in a sound field is useful in a number of acoustic signal processing problems. For instance, when performing dereverberation, an SDR estimate can be employed in an algorithm to suppress diffuse energy, which is detrimental to speech intelligibility [24, 36], while the direct sound and early reflections can be retained. Such an estimate can also be used to improve the accuracy of DOA estimation algorithms, such as those presented in Sect. 5.1, by discarding the inaccurate DOA estimates that are obtained when the sound field is highly diffuse. Finally, the SDR is related to the diffuseness, which is a key parameter in the description of spatial sound, for example in Directional Audio Coding (DirAC) [29].

In the spatial domain, SDR estimation has previously been accomplished using the coherence between a pair of omnidirectional microphones [43], or the coherence between a pair of first-order microphones [42]. However, spherical microphone arrays typically include significantly more than two microphones. We next present two methods in the SHD that take advantage of the availability of these additional microphone signals. We begin with a method based on the pseudointensity vectors introduced in Sect. 5.1.3, followed by a method based on the coherence between the eigenbeams.

5.2.1 Problem Formulation

Signal Models

Let the sound pressure $X(k, \mathbf{r})$ measured at a position $\mathbf{r}$ be modelled as[4] the sum of a directional signal X_{dir}, a diffuse signal X_{diff} and a sensor noise signal V:

$$X(k, \mathbf{r}) = X_{\text{dir}}(k, \mathbf{r}, \Omega_{\text{dir}}) + X_{\text{diff}}(k, \mathbf{r}) + V(k, \mathbf{r}). \tag{5.49}$$

In the following, we assume that the three signal components can be modelled by mutually uncorrelated complex Gaussian random variables with zero mean. The directional signal corresponds to the pressure due to a plane wave incident from a direction $\Omega_{\text{dir}} = (\theta_{\text{dir}}, \phi_{\text{dir}})$, where θ_{dir} represents inclination and ϕ_{dir} represents azimuth. The diffuse signal corresponds to the pressure due to an infinite number of independent plane waves, equal powers and uniformly distributed DOAs [20]. The powers of the directional and diffuse signals at an omnidirectional reference microphone $\mathcal{M}_{\text{ref}}$ (see the Appendix) are respectively denoted as $P_{\text{dir}}(k)$ and $P_{\text{diff}}(k)$.

When using spherical microphone arrays, it is convenient to work in the SHD [22, 30], instead of the spatial domain. We assume error-free spatial sampling, and refer the reader to Chap. 3 for information on spatial sampling and aliasing. By applying

[4] As in Sect. 5.1, the dependency on time is omitted for brevity.

the complex SHT, the spatial domain signal model in (5.49) can be expressed in the SHD as

$$X_{lm}(k) = X_{lm}^{\mathrm{dir}}(k, \Omega_{\mathrm{dir}}) + X_{lm}^{\mathrm{diff}}(k) + V_{lm}(k), \tag{5.50}$$

where the eigenbeams $X_{lm}(k)$, $X_{lm}^{\mathrm{dir}}(k)$, $X_{lm}^{\mathrm{diff}}(k)$ and $V_{lm}(k)$ respectively denote the SHTs of $X(k, \mathbf{r})$, $X_{\mathrm{dir}}(k, \mathbf{r})$, $X_{\mathrm{diff}}(k, \mathbf{r})$ and $V(k, \mathbf{r})$, as defined in (3.6).

In the SHD, the directional signal X_{lm}^{dir} is given by

$$X_{lm}^{\mathrm{dir}}(k, \Omega_{\mathrm{dir}}) = \sqrt{P_{\mathrm{dir}}(k)} A_{\mathrm{dir}}(k) 4\pi b_l(k) Y_{lm}^*(\Omega_{\mathrm{dir}}), \tag{5.51}$$

where $A_{\mathrm{dir}}(k)$ is a complex Gaussian random variable with zero mean and unit variance such that $\mathrm{E}\left\{|A_{\mathrm{dir}}(k)|^2\right\} = 1, \ \forall k$, and $Y_{lm}(\Omega_{\mathrm{dir}})$ is the complex spherical harmonic[5] of order l and degree m evaluated at an angle Ω_{dir}, as defined in (2.14). The frequency-dependent mode strength $b_l(k)$ captures the dependence of the eigenbeams on the array properties, and is discussed in Sect. 3.4.2.

The diffuse signal X_{lm}^{diff} can be expressed as

$$X_{lm}^{\mathrm{diff}}(k) = \sqrt{\frac{P_{\mathrm{diff}}(k)}{4\pi}} \int_{\Omega \in \mathcal{S}^2} A_{\mathrm{diff}}(k, \Omega) 4\pi b_l(k) Y_{lm}^*(\Omega) \mathrm{d}\Omega, \tag{5.52}$$

where $A_{\mathrm{diff}}(k, \Omega)$ are mutually uncorrelated complex Gaussian random variables with zero mean and unit variance such that $\mathrm{E}\left\{|A_{\mathrm{diff}}(k, \Omega)|^2\right\} = 1, \ \forall k, \Omega$ and

$$\mathrm{E}\left\{A_{\mathrm{diff}}(k, \Omega) A_{\mathrm{diff}}^*(k, \Omega')\right\} = \delta_{\Omega, \Omega'}, \tag{5.53}$$

δ denotes the Kronecker delta defined in (2.19), $\mathrm{E}\{\cdot\}$ denotes mathematical expectation, and the notation $\int_{\Omega \in \mathcal{S}^2} \mathrm{d}\Omega$ is used to denote compactly the solid angle $\int_{\phi=0}^{2\pi} \int_{\theta=0}^{\pi} \sin\theta \mathrm{d}\theta \mathrm{d}\phi$.

Using the relationship (5.74) between the zero-order eigenbeam $X_{00}(k)$ and the signal received at the reference microphone $\mathcal{M}_{\mathrm{ref}}$, as well as the expressions for the directional and diffuse signals in (5.51) and (5.52), it can be verified that the powers of these signals at $\mathcal{M}_{\mathrm{ref}}$ are respectively given by P_{dir} and P_{diff}.

Signal-to-Diffuse Ratio and Diffuseness

The SDR measures the ratio of the directional to diffuse signal power at a particular position in the sound field. The SDR Γ at $\mathcal{M}_{\mathrm{ref}}$ is given by

$$\Gamma(k) = \frac{\mathrm{E}\left\{|X_{00}^{\mathrm{dir}}(k, \Omega_{\mathrm{dir}})|^2\right\}}{\mathrm{E}\left\{|X_{00}^{\mathrm{diff}}(k)|^2\right\}} = \frac{P_{\mathrm{dir}}(k)}{P_{\mathrm{diff}}(k)}. \tag{5.54}$$

[5] As noted earlier in the chapter, if the real SHT is applied instead of the complex SHT, the complex spherical harmonics Y_{lm} used throughout this chapter should be replaced with the real spherical harmonics R_{lm}, as defined in Sect. 3.3.

The diffuseness Ψ of the sound field at $\mathcal{M}_{\text{ref}}$ is defined as the ratio of the directional signal power P_{dir} to the total signal power $P_{\text{dir}} + P_{\text{diff}}$, i.e.,

$$\Psi(k) = \frac{P_{\text{dir}}(k)}{P_{\text{dir}}(k) + P_{\text{diff}}(k)} = \frac{1}{1 + \Gamma(k)}. \tag{5.55}$$

The diffuseness takes values between 0 and 1; a value of 0 is obtained when $\Gamma(k) \to \infty$, corresponding to a purely directional field; a value of 1 is obtained when $\Gamma(k) = 0$, corresponding to a purely directional field; and a value of 0.5 is obtained when $\Gamma(k) = 1$, that is, when the directional and diffuse signals have equal power.

In Sects. 5.2.2 and 5.2.3, we describe two methods for estimating the SDR or diffuseness based on the eigenbeams $X_{lm}(k)$.

5.2.2 *Coefficient-of-Variation Method*

The *coefficient of variation* (CV) method proposed by Ahonen and Pulkki [1] exploits the temporal variation of the active intensity vector $\mathcal{I}$ to estimate the diffuseness $\Psi(k)$. In particular,

- in a purely diffuse field [11],

$$||\text{E}\{\mathcal{I}(k)\}|| = 0; \tag{5.56}$$

- in a purely directional field [7],

$$||\text{E}\{\mathcal{I}(k)\}|| = \text{E}\{||\mathcal{I}(k)||\}, \tag{5.57}$$

  where $||\cdot||$ denotes the 2-norm (Euclidian norm).

The CV method then estimates the diffuseness as

$$\Psi_{\text{CV}}(k) = \sqrt{1 - \frac{||\text{E}\{\mathcal{I}(k)\}||}{\text{E}\{||\mathcal{I}(k)||\}}}. \tag{5.58}$$

For a purely diffuse field, we have $\Psi_{\text{CV}} = 1$, while for a purely directional field, we have $\Psi_{\text{CV}} = \sqrt{1 - 1/1} = 0$, as desired. The square root is applied in order to approximate more accurately the diffuseness [7] defined in (5.55).

As discussed in Sect. 5.1.3, the intensity vector can be approximated (to within a scaling factor) by the pseudointensity vector, which is computed using a linear combination of zero- and first-order eigenbeams. The CV method can then be applied to the pseudointensity vectors, and is then referred to as the *modified* CV method [16] to distinguish it from the original method proposed by Ahonen & Pulkki. As noted in Sect. 5.1.3, while the intensity vector can be measured directly using acoustic vector sensors, the pseudointensity vector is more robust to noise, due to the fact that it is estimated using all available microphone signals.

5.2.3 Coherence-Based Method

The coherence-based method proposed in [16] exploits the fact that the coherence between eigenbeams is an increasing function of the SDR. In this section, we derive expressions for the coherence in both purely directional and purely diffuse fields, and then derive a relationship between the SDR and the coherence in a mixed field.

The complex coherence $\gamma_{lm,l'm'}(k)$ between the eigenbeams $X_{lm}(k)$ and $X_{l'm'}(k)$ is defined as

$$\gamma_{lm,l'm'}(k) = \frac{\Phi_{lm,l'm'}(k)}{\sqrt{\Phi_{lm,lm}(k)}\sqrt{\Phi_{l'm',l'm'}(k)}}, \tag{5.59}$$

where the power spectral densities (PSDs) $\Phi_{lm,l'm'}$ are given by

$$\Phi_{lm,l'm'}(k) = \mathrm{E}\left\{X_{lm}(k)X^*_{l'm'}(k)\right\}. \tag{5.60}$$

In a purely directional field, using (5.60) and (5.51), and the fact that $\mathrm{E}\left\{|A_{\mathrm{dir}}(k)|^2\right\} = 1$, the PSD $\Phi^{\mathrm{dir}}_{lm,l'm'}(k)$ is expressed as

$$\Phi^{\mathrm{dir}}_{lm,l'm'}(k) = \mathrm{E}\left\{X^{\mathrm{dir}}_{lm}(k)\left(X^{\mathrm{dir}}_{l'm'}(k)\right)^*\right\} \tag{5.61a}$$

$$= P_{\mathrm{dir}}(k)(4\pi)^2 b_l(k)b^*_{l'}(k)Y^*_{lm}(\Omega_{\mathrm{dir}})Y_{l'm'}(\Omega_{\mathrm{dir}}). \tag{5.61b}$$

By substituting (5.61b) into (5.59), we obtain the directional field coherence $\gamma^{\mathrm{dir}}_{lm,l'm'}$:

$$\gamma^{\mathrm{dir}}_{lm,l'm'}(k) = \frac{b_l(k)b^*_{l'}(k)Y^*_{lm}(\Omega_{\mathrm{dir}})Y_{l'm'}(\Omega_{\mathrm{dir}})}{|b_l(k)b^*_{l'}(k)Y^*_{lm}(\Omega_{\mathrm{dir}})Y_{l'm'}(\Omega_{\mathrm{dir}})|}. \tag{5.62}$$

In a purely directional field, the coherence therefore has unit magnitude.

In a purely diffuse field, using (5.60) and (5.52), the PSD $\Phi^{\mathrm{diff}}_{lm,l'm'}(k)$ is expressed as

$$\Phi^{\mathrm{diff}}_{lm,l'm'}(k) = \mathrm{E}\left\{X^{\mathrm{diff}}_{lm}(k)\left(X^{\mathrm{diff}}_{l'm'}(k)\right)^*\right\} \tag{5.63a}$$

$$= P_{\mathrm{diff}}(k)4\pi\,\mathrm{E}\left\{\int_{\Omega\in\mathcal{S}^2} A_{\mathrm{diff}}(k,\Omega)b_l(k)Y^*_{lm}(\Omega)\mathrm{d}\Omega \times \int_{\Omega'\in\mathcal{S}^2} A^*_{\mathrm{diff}}(k,\Omega')b^*_{l'}(k)Y_{l'm'}(\Omega')\mathrm{d}\Omega'\right\}. \tag{5.63b}$$

Using (5.53), the orthonormality of the spherical harmonics (2.18), and the fact that $\mathrm{E}\left\{|A_{\mathrm{diff}}(k,\Omega)|^2\right\} = 1$, (5.63b) can be simplified to

$$\Phi^{\text{diff}}_{lm,l'm'}(k) = P_{\text{diff}}(k)4\pi \int_{\Omega \in \mathcal{S}^2} b_l(k)b^*_{l'}(k)Y^*_{lm}(\Omega)Y_{l'm'}(\Omega)\mathrm{d}\Omega \tag{5.64a}$$

$$= P_{\text{diff}}(k)4\pi b_l(k)b^*_{l'}(k)\delta_{l,l'}\delta_{m,m'}. \tag{5.64b}$$

By substituting (5.64b) into (5.59), we obtain the diffuse field coherence $\gamma^{\text{diff}}_{lm,l'm'}$:

$$\gamma^{\text{diff}}_{lm,l'm'}(k) = \frac{b_l(k)b^*_{l'}(k)}{|b_l(k)b_{l'}(k)|}\delta_{l,l'}\delta_{m,m'} \tag{5.65a}$$

$$= \begin{cases} \frac{b_l(k)b^*_{l'}(k)}{|b_l(k)b_{l'}(k)|}, & \text{if } (l,m) = (l',m'), \\ 0, & \text{otherwise.} \end{cases} \tag{5.65b}$$

In a purely diffuse field, the coherence between a non-identical pair of eigenbeams is therefore zero.

We assume the sensor noise V is spatially incoherent and has equal power $P_{\text{N}}(k)$ at each of the Q microphones uniformly distributed on the sphere. The noise eigenbeams $V_{lm}(k)$ are therefore also incoherent across l and m, and the noise PSD $\Phi^{\text{N}}_{lm,l'm'}$ is therefore given by [46, Eq. 7.31]

$$\Phi^{\text{N}}_{lm,l'm'}(k) = \text{E}\left\{V_{lm}(k)V^*_{l'm'}(k)\right\} \tag{5.66a}$$

$$= P_{\text{N}}(k)\frac{4\pi}{Q}\delta_{l,l'}\delta_{m,m'} \tag{5.66b}$$

$$= \begin{cases} P_{\text{N}}(k)\frac{4\pi}{Q}, & \text{if } (l,m) = (l',m'), \\ 0, & \text{otherwise.} \end{cases} \tag{5.66c}$$

In a mixed sound field, the directional, diffuse and noise signals are all present. We assume that they are mutually uncorrelated; hence, the PSD $\Phi_{lm,l'm'}$ is equal to the sum of the PSDs in (5.61b), (5.64b) and (5.66c):

$$\Phi_{lm,l'm'}(k) = \text{E}\left\{X_{lm}(k)X^*_{l'm'}(k)\right\} \tag{5.67a}$$

$$= \Phi^{\text{dir}}_{lm,l'm'}(k) + \Phi^{\text{diff}}_{lm,l'm'}(k) + \Phi^{\text{N}}_{lm,l'm'}(k). \tag{5.67b}$$

We define the *noiseless* coherence as

$$\mathring{\gamma}_{lm,l'm'}(k) = \frac{\mathring{\Phi}_{lm,l'm'}(k)}{\sqrt{\mathring{\Phi}_{lm,lm}(k)}\sqrt{\mathring{\Phi}_{l'm',l'm'}(k)}}, \tag{5.68}$$

where the noiseless PSD $\mathring{\Phi}_{lm,l'm'}(k)$ is defined as $\mathring{\Phi}_{lm,l'm'}(k) = \Phi^{\text{dir}}_{lm,l'm'}(k) + \Phi^{\text{diff}}_{lm,l'm'}(k)$. Using (5.61b) and (5.64b), the noiseless PSD can be expressed as

$$\mathring{\Phi}_{lm,l'm'}(k) = P_{\text{dir}}(k)(4\pi)^2 b_l(k)b^*_{l'}(k)Y^*_{lm}(\Omega_{\text{dir}})Y_{l'm'}(\Omega_{\text{dir}}) + P_{\text{diff}}(k)4\pi b_l(k)b^*_{l'}(k)\delta_{l,l'}\delta_{m,m'}. \tag{5.69}$$

By substituting (5.69) in (5.68), and using (5.54), it can be shown that [16, 42]

$$\mathring{\gamma}_{lm,l'm'}(k) = \frac{\Gamma(k)\gamma^{\text{dir}}_{lm,l'm'}(k)c_{lm}c_{l'm'}}{\sqrt{\Gamma^2(k)c^2_{lm}c^2_{l'm'} + \Gamma(k)(c^2_{lm} + c^2_{l'm'}) + 1}}, \tag{5.70}$$

where $c_{lm} = \sqrt{4\pi}|Y_{lm}(\Omega_{\text{dir}})|$.

The noiseless PSD $\mathring{\Phi}_{lm,l'm'}(k)$ cannot be directly observed; however, with sufficient time averaging, the noise cross PSD $\Phi^{\text{N}}_{lm,l'm'}$ $[(l,m) \neq (l',m')]$ will average to 0. The noiseless auto PSD $\mathring{\Phi}_{lm,lm}$ can be estimated using an estimate of the noise power $P_{\text{N}}(k)$:

$$\mathring{\Phi}_{lm,lm}(k) = \Phi_{lm,lm}(k) - \Phi^{\text{N}}_{lm,lm}(k) \tag{5.71a}$$

$$= \Phi_{lm,lm}(k) - P_{\text{N}}(k)\frac{4\pi}{Q}. \tag{5.71b}$$

The SDR is determined by solving for $\Gamma(k)$ in (5.70), as in [42]:

$$\hat{\Gamma}_{lm,l'm'}(k) = \frac{G(\Omega_{\text{dir}}) + \sqrt{G^2(\Omega_{\text{dir}}) + 4\left(\left|\mathring{\gamma}_{lm,l'm'}(k)\right|^{-2} - 1\right)}}{2c_{lm}(\Omega_{\text{dir}})c_{l'm'}(\Omega_{\text{dir}})\left(\left|\mathring{\gamma}_{lm,l'm'}(k)\right|^{-2} - 1\right)}, \tag{5.72}$$

where we have defined

$$G(\Omega_{\text{dir}}) = \frac{c_{lm}(\Omega_{\text{dir}})}{c_{l'm'}(\Omega_{\text{dir}})} + \frac{c_{l'm'}(\Omega_{\text{dir}})}{c_{lm}(\Omega_{\text{dir}})}. \tag{5.73}$$

The DOA Ω_{dir} must be estimated in order to compute the coefficients c_{lm}; for this purpose, one of the algorithms presented in Sect. 5.1 can be used.

The performance of this coherence-based method depends on which pair of non-identical eigenbeams is chosen and on the DOA Ω_{dir}. One way of improving the performance is to compute the coherence between *all* pairs of non-identical eigenbeams, instead of the coherence between a single pair of eigenbeams. An SDR estimate with lower variance is then obtained by computing the weighted average of the coherences, as in [16] where the weights are given by the geometric average of the directional signal-to-noise ratio of the eigenbeams involved.

5.2.4 Results

In this section, we provide some illustrative results from the modified CV method and the coherence-based method. The signals received by a rigid spherical microphone array of radius $r = 4.2$ cm were simulated directly in the SHD. The directional signals consisted of complex white Gaussian noise, originating from a direction $(\theta_{\text{dir}}, \phi_{\text{dir}}) = (90°, 0°)$. The diffuse signals were generated using 1000 plane waves with uniform DOAs and random phase. The power of the diffuse signals was set according to the desired SDR. The noise signals consisted of complex white Gaussian noise; its power was set such that a directional-signal-to-noise ratio of 25 dB was obtained at the arbitrarily chosen reference microphone $\mathcal{M}_{\text{ref}}$. The received signals were processed in the STFT domain at a sampling frequency of 8 kHz, with a frame length of 16 ms and a 50 % overlap between successive frames.

Figure 5.4 shows the azimuths of the pseudointensity vectors obtained at a frequency of 1 kHz in three sound fields with different SDRs. For clarity, only the pseudointensity vectors corresponding to the first 100 time frames are shown. As shown in Fig. 5.4a, in a purely directional field the pseudointensity vectors all point in the opposite direction to the directional source, as expected (see Sect. 5.1.3).

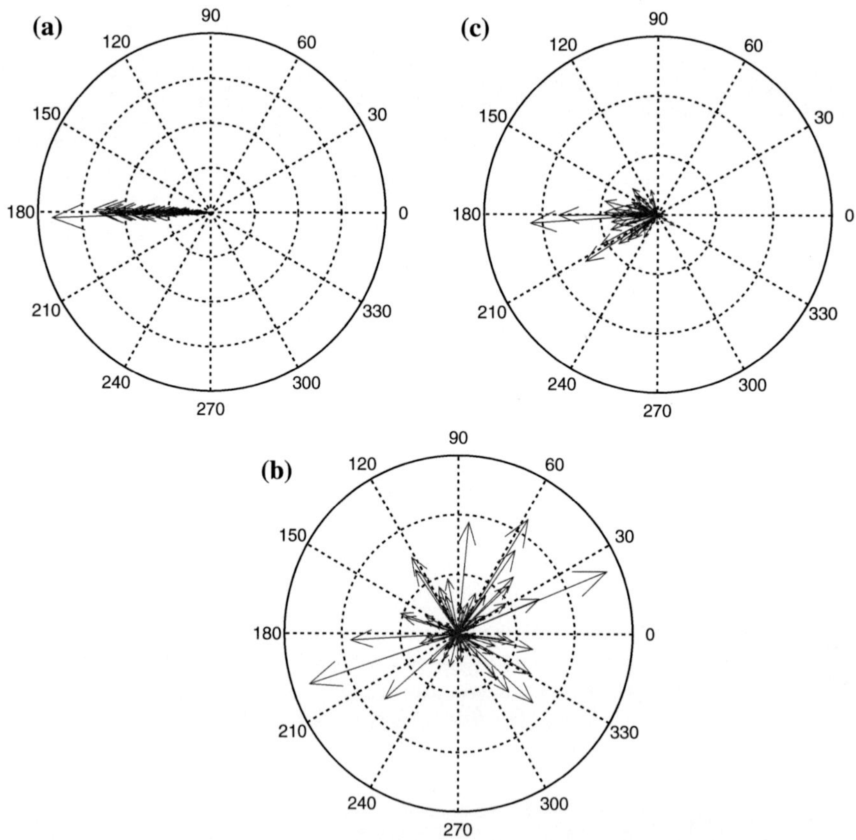

Fig. 5.4 Azimuths (in degrees) of the pseudointensity vectors obtained at a frequency of 1 kHz in three different sound fields with varying SDRs. The directional field was due to a plane wave originating from an azimuth of 0°; the pseudointensity vector points in the opposite direction to the sound source. **a** Purely directional field (infinite SDR), **b** Purely diffuse field (SDR of 0), **c** Mixed field (SDR of 1)

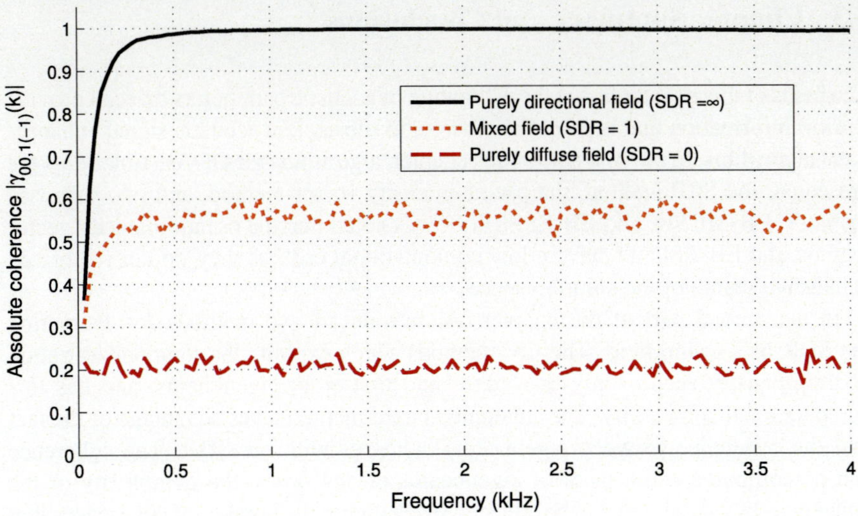

Fig. 5.5 Absolute coherence $|\gamma_{00,1(-1)}(k)|$ obtained as a function of frequency in three different sound fields with varying SDRs. The absolute coherence values were computed for each time-frequency bin, and averaged over 5 s of data

As shown in Fig. 5.4b, in a purely diffuse field the direction of the pseudointensity vectors is random, reflecting the fact that the diffuse field is composed of many plane waves with different DOAs. Finally, as shown in Fig. 5.4c, in a mixed sound field with an SDR of 0 dB at $\mathcal{M}_{\text{ref}}$ most of the pseudointensity vectors point in the opposite direction to the directional source, but there is more variance than in the purely directional case. From these results, it is clear that the temporal variation of the pseudointensity vectors can be exploited for SDR estimation.

The absolute coherence $|\gamma_{00,1(-1)}(k)|$ between the eigenbeams $X_{00}(k)$ and $X_{1(-1)}(k)$ for these same three sound fields is shown in Fig. 5.5 as a function of frequency. The absolute coherence values were averaged over 5 s of data. The coherence was computed using (5.59); the expectations were approximated using moving averages over 20 time frames. As expected, the coherence is highest in a purely directional field, and lowest in a purely diffuse field. The non-zero coherence in a purely diffuse field is due to the finite time averaging involved in computing the expectations. The low coherence at low frequencies (below 250 Hz) in directional fields is due to the low directional signal-to-noise ratio at these frequencies. Further illustrative results relating to coherence-based SDR estimation can be found in [16].

5.3 Chapter Summary and Conclusions

The focus of this chapter lies in the estimation of acoustic parameters that can provide a priori information that is potentially useful to subsequent acoustic signal enhancement algorithms. In the first part of this chapter, algorithms for DOA estimation were presented: the SRP method, the pseudointensity vector method, and two subspace methods, EB-MUSIC and EB-ESPRIT. It was noted that the pseudointensity vector method and EB-ESPRIT have a low computational cost, as they do not require an exhaustive search of the solution space.

In the second part of the chapter, we introduced two methods for estimating the SDR of a sound field. The CV method, which exploits the temporal variation of the intensity vector, only uses zero- and first-order eigenbeams, and has low computational complexity. The coherence-based method takes advantage of the fact that the coherence between eigenbeams increases with the SDR. The coherence can be computed using pairs of eigenbeams of any order; the complexity of the coherence-based method can be adjusted by changing the number of coherences that are computed. The best way of estimating the coherence using all of the available eigenbeams in a computationally efficient way remains an open question at this time.

Appendix: Relationship Between the Zero-Order Eigenbeam and the Omnidirectional Reference Microphone Signal

Property 5.1 *Let $P_{lm}(k)$ denote the SHT, as defined in (3.6), of the spatial domain sound pressure $P(k, \mathbf{r})$, where $\mathbf{r}$ denotes the position (in spherical coordinates) with respect to the centre of a spherical microphone array with mode strength $b_l(k)$. Let $P_{\mathcal{M}_{ref}}(k)$ denote the sound pressure which would be measured using an omnidirectional microphone $\mathcal{M}_{ref}$ at a position corresponding to the centre of the sphere, i.e., at the origin of the spherical coordinate system; $P_{\mathcal{M}_{ref}}(k)$ is then related to the zero-order eigenbeam $P_{00}(k)$ via the relationship*[6]

$$P_{\mathcal{M}_{ref}}(k) = \frac{P_{00}(k)}{\sqrt{4\pi}\, b_0(k)}. \tag{5.74}$$

Proof We assume, without loss of generality,[7] that the sound field is composed of a single unit amplitude spherical wave incident from a point source at a position $\mathbf{r}_s = (r_s, \Omega_s)$.

---

[6]It should be noted that this relationship is dependent upon the chosen mode strength definition (see Sect. 3.4.2). If a 4π factor is included in $b_l(k)$, as in [31], the relationship becomes $P_{\mathcal{M}_{ref}}(k) = \sqrt{4\pi}\frac{P_{00}(k)}{b_0(k)}$.

[7]The operations involved in the proof are linear, and the proof therefore holds for any number of spherical waves.

In the absence of the sphere, the sound pressure measured at the origin of the spherical coordinate system due to a single spherical wave incident from a point source at a position $\mathbf{r}_s = (r_s, \Omega_s)$ is given by (4.7), i.e.,

$$P_{\mathcal{M}_{\text{ref}}}(k) = \frac{e^{-ik||\mathbf{r}_s||}}{4\pi\,||r_s||} \tag{5.75a}$$

$$= \frac{e^{-ikr_s}}{4\pi r_s}. \tag{5.75b}$$

In the spatial domain, the sound pressure $P(k, \mathbf{r})$ at a position $\mathbf{r}$ due to the spherical wave is given by (4.10), and can be written using (2.23) as

$$P(k, \mathbf{r}) = -ik \sum_{l=0}^{\infty} i^{-l} b_l(k) h_l^{(2)}(kr_s) \sum_{m=-l}^{l} Y_{lm}^*(\Omega_s) Y_{lm}(\Omega), \tag{5.76}$$

where $h_l^{(2)}$ is the spherical Hankel function of the second kind and of order l. From the definition of the SHT (3.7), $P_{00}(k)$ is given by

$$P_{00}(k) = \int_{\Omega \in S^2} P(k, \mathbf{r}) Y_{00}^*(\Omega) \mathrm{d}\Omega. \tag{5.77}$$

By substituting (5.76) into (5.77), we find

$$P_{00}(k) = \int_{\Omega \in S^2} -ik \sum_{l=0}^{\infty} i^{-l} b_l(k) h_l^{(2)}(kr_s) \sum_{m=-l}^{l} Y_{lm}^*(\Omega_s) Y_{lm}(\Omega) Y_{00}^*(\Omega) \mathrm{d}\Omega. \tag{5.78}$$

Using the orthonormality of the spherical harmonics (2.18) and the fact that $Y_{00}(\cdot) = 1/\sqrt{4\pi}$, we can simplify (5.78) to

$$P_{00}(k) = -ikb_0(k) h_0^{(2)}(kr_s) Y_{00}^*(\Omega_s) \tag{5.79a}$$

$$= -\frac{ik}{\sqrt{4\pi}} b_0(k) h_0^{(2)}(kr_s). \tag{5.79b}$$

Finally, using the fact that $h_0^{(2)}(x) = \frac{-e^{-ix}}{ix}$ [46, Eq. 6.62] and (5.75b), we can simplify (5.79b) to

$$P_{00}(k) = \frac{ik}{\sqrt{4\pi}} b_0(k) \frac{e^{-ikr_s}}{ikr_s} \tag{5.80a}$$

$$= \sqrt{4\pi}\, b_0(k) \frac{e^{-ikr_s}}{4\pi r_s} \tag{5.80b}$$

$$= \sqrt{4\pi}\, b_0(k)\, P_{\mathcal{M}_{\text{ref}}}(k), \tag{5.80c}$$

and therefore Property 5.1 holds.

References

1. Ahonen, J., Pulkki, V.: Diffuseness estimation using temporal variation of intensity vectors. In: Proceedings of the IEEE Workshop on Applications of Signal Processing to Audio and Acoustics, pp. 285–288 (2009)
2. Ahonen, J., Pulkki, V., Lokki, T.: Teleconference application and B-format microphone array for directional audio coding. In: Proceedings of the AES 30th International Conference (2007)
3. Berge, S., Barrett, N.: High angular resolution planewave expansion. In: Proceedings of the 2nd International Symposium on Ambisonics and Spherical Acoustics (2010)
4. Chapman, M., Ritsch, W., Musil, T., Zmölnig, I., Pomberger, H., Zotter, F., Sontacchi, A.: A standard for interchange of Ambisonic signal sets. In: Proceedings of the Ambisonics Symposium (2009)
5. Crocker, M.J. (ed.): Handbook of Acoustics. Wiley-Interscience, New York (1998)
6. Crocker, M.J., Jacobsen, F.: Sound intensity. In: Crocker [5], chap. 106, pp. 1327–1340
7. Del Galdo, G., Taseska, M., Thiergart, O., Ahonen, J., Pulkki, V.: The diffuse sound field in energetic analysis. J. Acoust. Soc. Am. **131**(3), 2141–2151 (2012)
8. Evers, C., Moore, A.H., Naylor, P.A.: Multiple source localisation in the spherical harmonic domain. In: Proceedings of the International Workshop Acoust. Signal Enhancement (IWAENC). Nice, France (2014)
9. Fletcher, H., Munson, W.A.: Loudness, its definition, measurement and calculation. J. Acoust. Soc. Am. **5**(1), 82–108 (1933)
10. Hahn, W.: Optimum signal processing for passive sonar range and bearing estimation. J. Acoust. Soc. Am. **58**(1), 201–207 (1975)
11. Jacobsen, F.: Active and reactive sound intensity in a reverberant sound field. J. Sound Vib. **143**(2), 231–240 (1990). doi:10.1016/0022-460X(90)90952-V
12. Jarrett, D.P.: Spherical Microphone array Impulse Response (SMIR) generator. http://www.ee.ic.ac.uk/sap/smirgen/
13. Jarrett, D.P.: Spherical microphone array processing for acoustic parameter estimation and signal enhancement. Ph.D. thesis, Imperial College London (2013)
14. Jarrett, D.P., Habets, E.A.P., Naylor, P.A.: 3D source localization in the spherical harmonic domain using a pseudointensity vector. In: Proceedings of the European Signal Processing Conference (EUSIPCO), pp. 442–446. Aalborg, Denmark (2010)
15. Jarrett, D.P., Habets, E.A.P., Naylor, P.A.: Eigenbeam-based acoustic source tracking in noisy reverberant environments. In: Proceedings of the Asilomar Conference on Signals, Systems and Computers, pp. 576–580. Pacific Grove, CA, USA (2010)
16. Jarrett, D.P., Thiergart, O., Habets, E.A.P., Naylor, P.A.: Coherence-based diffuseness estimation in the spherical harmonic domain. In: Proceedings of the IEEE Convention of Electrical & Electronics Engineers in Israel (IEEEI). Eilat, Israel (2012)
17. Khaykin, D., Rafaely, B.: Signal decorrelation by spatial smoothing for spherical microphone arrays. In: Proceedings of the IEEE Convention of Electrical & Electronics Engineers in Israel (IEEEI), pp. 270–274 (2008)
18. Khaykin, D., Rafaely, B.: Coherent signals direction-of-arrival estimation using a spherical microphone array: Frequency smoothing approach. In: Proceedings of the IEEE Workshop on Applications of Signal Processing to Audio and Acoustics, pp. 221–224 (2009). doi:10.1109/ASPAA.2009.5346492
19. Knapp, C., Carter, G.: The generalized correlation method for estimation of time delay. IEEE Trans. Acoust. Speech, Signal Process. **24**(4), 320–327 (1976)
20. Kuttruff, H.: Room Acoustics, 4th edn. Taylor & Francis, London (2000)
21. Merimaa, J.: Analysis, synthesis, and perception of spatial sound âĂŞ binaural localization modeling and multichannel loudspeaker reproduction. Ph.D. thesis, Helsinki University of Technology (2006)
22. Meyer, J., Elko, G.: A highly scalable spherical microphone array based on an orthonormal decomposition of the soundfield. In: Proceedings of the IEEE International Conference on Acoustics, Speech and Signal Processing (ICASSP), vol. 2, pp. 1781–1784 (2002)

23. Moore, A.H., Evers, C., Naylor, P.A., Alon, D.L., Rafaely, B.: Direction of arrival estimation using pseudo-intensity vectors with direct-path dominance test. In: Proceedings of the European Signal Processing Conference (EUSIPCO) (2015)
24. Nábělek, A.K., Mason, D.: Effect of noise and reverberation on binaural and monaural word identification by subjects with various audiograms. J Speech and Hear. Res. **24**, 375–383 (1981)
25. Nadiri, O., Rafaely, B.: Localization of multiple speakers under high reverberation using a spherical microphone array and the direct-path dominance test **22**(10), 1494–1505 (2014)
26. Nehorai, A., Paldi, E.: Acoustic vector-sensor array processing. IEEE Trans. Signal Process. **42**(9), 2481–2491 (1994). doi:10.1109/78.317869
27. Peled, Y., Rafaely, B.: Linearly constrained minimum variance method for spherical microphone arrays in a coherent environment. In: Proceedings of the Hands-Free Speech Communication and Microphone Arrays (HSCMA), pp. 86–91 (2011). doi:10.1109/HSCMA.2011.5942416
28. Pierce, A.D.: Acoustics: An Introduction to Its Physical Principles and Applications. Acoustical Society of America (1991)
29. Pulkki, V.: Spatial sound reproduction with directional audio coding. J. Audio Eng. Soc. **55**(6), 503–516 (2007)
30. Rafaely, B.: Plane-wave decomposition of the pressure on a sphere by spherical convolution. J. Acoust. Soc. Am. **116**(4), 2149–2157 (2004)
31. Rafaely, B.: Analysis and design of spherical microphone arrays. IEEE Trans. Speech Audio Process. **13**(1), 135–143 (2005). doi:10.1109/TSA.2004.839244
32. Rafaely, B., Peled, Y., Agmon, M., Khaykin, D., Fisher, E.: Spherical microphone array beamforming. In: I. Cohen, J. Benesty, S. Gannot (eds.) Speech Processing in Modern Communication: Challenges and Perspectives, chap. 11. Springer, Berlin (2010)
33. Roy, R., Kailath, T.: ESPRIT-estimation of signal parameters via rotational invariance techniques. IEEE Trans. Acoust. Speech, Signal Process. **37**, 984–995 (1989)
34. Schmidt, R.: A signal subspace approach to multiple emitter location and spectral estimation. Ph.D. thesis, Stanford University, Stanford, CA. (1981)
35. Schmidt, R.O.: Multiple emitter location and signal parameter estimation. IEEE Trans. Antennas Propag. **34**(3), 276–280 (1986)
36. Steinberg, J.C.: Effects of distortion upon the recognition of speech sounds. J. Acoust. Soc. Am. **1**, 35–35 (1929)
37. Sun, H., Mabande, E., Kowalczyk, K., Kellermann, W.: Localization of distinct reflections in rooms using spherical microphone array eigenbeam processing. J. Acoust. Soc. Am. **131**(4), 2828–2840 (2012)
38. Sun, H., Teutsch, H., Mabande, E., Kellermann, W.: Robust localization of multiple sources in reverberant environments using EB-ESPRIT with spherical microphone arrays. In: Proceedings of the IEEE International Conference on Acoustics, Speech and Signal Processing (ICASSP), pp. 117–120. Prague, Czech Republic (2011)
39. Teutsch, H.: Wavefield decomposition using microphone arrays and its application to acoustic scene analysis. Ph.D. thesis, Friedrich-Alexander Universität Erlangen-Nürnberg (2005)
40. Teutsch, H., Kellermann, W.: Eigen-beam processing for direction-of-arrival estimation using spherical apertures. In: Proceedings of the Joint Workshop on Hands-Free Speech Communication and Microphone Arrays. Piscataway, New Jersey, USA (2005)
41. Teutsch, H., Kellermann, W.: Detection and localization of multiple wideband acoustic sources based on wavefield decomposition using spherical apertures. In: Proceedings of the IEEE International Conference on Acoustics, Speech and Signal Processing (ICASSP), pp. 5276–5279 (2008). doi:10.1109/ICASSP.2008.4518850
42. Thiergart, O., Del Galdo, G., Habets, E.A.P.: On the spatial coherence in mixed sound fields and its application to signal-to-diffuse ratio estimation. J. Acoust. Soc. Am. **132**(4), 2337–2346 (2012)
43. Thiergart, O., Del Galdo, G., Habets, E.A.P.: Signal-to-reverberant ratio estimation based on the complex spatial coherence between omnidirectional microphones. In: Proceedings of the IEEE International Conference on Acoustics, Speech and Signal Processing (ICASSP), pp. 309–312 (2012)

44. Thiergart, O., Habets, E.A.P.: Robust direction-of-arrival estimation of two simultaneous plane waves from a B-format signal. In: Proceedings of the IEEE Convention of Electrical & Electronics Engineers in Israel (IEEEI). Eilat, Israel (2012)
45. van Trees, H.L.: Optimum Array Processing. Detection, Estimation and Modulation Theory. Wiley, New York (2002)
46. Williams, E.G.: Fourier Acoustics: Sound Radiation and Nearfield Acoustical Holography, 1st edn. Academic Press, London (1999)

Chapter 6
Signal-Independent Array Processing

The process of combining signals acquired by a microphone array in order to 'focus' on a signal in a specific direction is known as beamforming or spatial filtering. We present in this chapter a number of such beamforming methods that are specifically controlled by weights dependent only on the direction of arrival (DOA) of the desired source. They are otherwise signal-independent such that they do not depend on the statistics of the desired or noise signals. We derive maximum directivity and maximum white noise gain beamformers that establish performance bounds for spherical harmonic domain (SHD) beamformers. Because the weights of these beamformers are given by simple expressions, they present the advantages of being straightforward to implement and of having low computational complexity.

6.1 Signal Model

The sound pressure P captured at a position $\mathbf{r} = (r, \Omega) = (r, \theta, \phi)$ (in spherical coordinates, where θ denotes the inclination and ϕ denotes the azimuth) on a spherical microphone array of radius r is commonly expressed as the sum of a desired signal X and a noise signal V [12, 15]. In the spatial domain, the signal model is expressed as

$$P(k, \mathbf{r}) = X(k, \mathbf{r}) + V(k, \mathbf{r}), \tag{6.1}$$

D. Jarrett et al., *Theory and Applications of Spherical Microphone Array Processing*, Springer Topics in Signal Processing 9,
DOI 10.1007/978-3-319-42211-4_6

where k denotes the wavenumber.[1] The desired signal X is assumed to be spatially coherent, while the noise signal V models background noise or sensor noise, for example, and may be spatially incoherent, coherent or partially coherent.

When using spherical microphone arrays, it is convenient to work in the SHD [1, 17]. In this chapter, we assume error-free spatial sampling by Q microphones at positions $\mathbf{r}_q = (r, \Omega_q), q \in \{1, \ldots, Q\}$, and refer the reader to Chap. 3 for information on spatial sampling and aliasing. By applying the complex spherical harmonic transform (SHT) to the signal model in (6.1), we obtain the SHD signal model

$$P_{lm}(k) = X_{lm}(k) + V_{lm}(k), \tag{6.2}$$

where $P_{lm}(k)$, $X_{lm}(k)$ and $V_{lm}(k)$ are respectively the spherical harmonic transforms of the spatial domain signals $P(k, \mathbf{r}_q)$, $X(k, \mathbf{r}_q)$ and $V(k, \mathbf{r}_q)$, as defined in (3.6), and are referred to as *eigenbeams* to reflect the fact that the spherical harmonics are eigensolutions of the wave equation in spherical coordinates [26]. The order and degree of the spherical harmonics are respectively denoted as l and m.

By combining the eigenbeams $P_{lm}(k)$ in a particular way, the noise V can be suppressed and the desired signal X can be extracted from the noisy mixture P. This is accomplished using a spatio-temporal filter or *beamformer*. In the spatial domain, the output of a beamformer is obtained as the weighted sum of the pressure signals at each of the microphones [3, 4]; in the SHD, the beamformer output is given by a weighted sum of the eigenbeams $P_{lm}(k)$ [14, 21]. The output of an Lth-order SHD beamformer can thus be expressed as [21, Eq. 12][2]

$$Z(k) = \sum_{l=0}^{L} \sum_{m=-l}^{l} W_{lm}^{*}(k) P_{lm}(k), \tag{6.3}$$

where $W_{lm}(k)$ denotes the beamformer weights and $(\cdot)^*$ denotes the complex conjugate.

Beamformers can either be signal-independent (fixed) or signal-dependent; their weights are chosen in order to achieve specific performance objectives. Signal-independent beamformers apply a constraint to a specific steering direction and optimize the beamformer weights with respect to array performance measures such as the white noise gain (WNG) and directivity. They can also, more generally, attempt to achieve a specific spatial response in all directions by minimizing the difference between the beamformer's spatial response and the desired spatial response, according to some distance measure (see [6, Sects. 8.3 and 8.4] for examples). Signal-dependent beamformers optimize the weights taking into account characteristics of

[1]The dependency on time is omitted for brevity. In practice, the signals acquired using a spherical microphone array are usually processed in the short-time Fourier transform domain, as explained in Sect. 3.1, where the discrete frequency index is denoted by ν.

[2]We use the complex conjugate weights W_{lm}^{*} rather than the weights W_{lm}; this notational convention originates in the spatial domain [30].

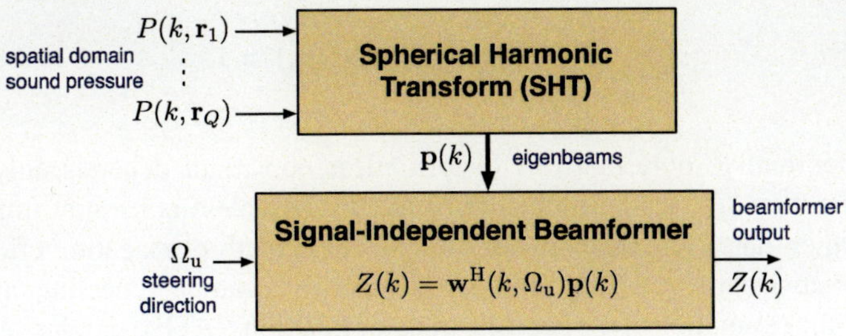

Fig. 6.1 Block diagram of a signal-independent beamformer

the desired signal and noise. In this chapter, we will discuss signal-independent beamformers and later address signal-dependent beamformers in Chap. 7.

A block diagram of a signal-independent beamformer is shown in Fig. 6.1. We begin by capturing the sound pressure signals $P(k, \mathbf{r}_q)$ at microphones $q \in \{1, \ldots, Q\}$, and applying the SHT to obtain the SHD sound pressure signals, the *eigenbeams* $P_{lm}(k)$, gathered together to form a vector $\mathbf{p}(k)$. The output $Z(k)$ of the beamformer is obtained by taking the weighted sum of these eigenbeams, where the weights $W_{lm}(k, \Omega_{\mathrm{u}})$ depend only on the steering direction Ω_{u} and do not otherwise depend on the sound pressure signals P.

The signal-independent beamformers presented in this chapter are designed assuming anechoic conditions with a single active sound source, though these assumptions are unlikely to be valid in practical use scenarios. Depending on the distance between this source and the array, the desired signal is either assumed to consist of a plane wave or a spherical wave. Under farfield conditions, the eigenbeams of a unit amplitude plane wave incident from a direction Ω_{s} are given by (3.22a). The SHD sound pressure $X_{lm}(k, \Omega_{\mathrm{s}})$ related to a plane wave with power $P_{\mathrm{pw}}(k)$ can then be written as [18, 20, 26]

$$X_{lm}(k, \Omega_{\mathrm{s}}) = \sqrt{P_{\mathrm{pw}}(k)} b_l(k) Y^*_{lm}(\Omega_{\mathrm{s}}), \tag{6.4}$$

where $Y_{lm}(\Omega_{\mathrm{s}})$ denotes the complex spherical harmonic[3] of order l and degree m evaluated at an angle Ω_{s}, as defined in (2.14), and the mode strength $b_l(k)$ captures the eigenbeams' dependence on the array properties, such as microphone type or array configuration, and is discussed in more detail in Sect. 3.4.2.

All the beamformers designed in this chapter seek to suppress the noise while maintaining a *distortionless constraint* on the signal originating from the steering direction Ω_{u}. This constraint is expressed as

[3]If the real SHT is applied instead of the complex SHT, the complex spherical harmonics Y_{lm} used throughout this chapter should be replaced with the real spherical harmonics R_{lm}, as defined in Sect. 3.3.

$$\sum_{l=0}^{L} \sum_{m=-l}^{l} W_{lm}^{*}(k) b_l(k) Y_{lm}^{*}(\Omega_{\mathrm{u}}) = 1. \tag{6.5}$$

It is important to note that this distortionless constraint depends only on the steering direction Ω_{u}. It is different from the distortionless constraint imposed in Chap. 7, which takes into account the complex multipath propagation effects of a reverberant environment. Using the constraint in (6.5) can be appealing, as it does not require the estimation of the acoustic transfer functions (ATFs) or relative transfer functions, however this comes at the expense of sensitivity to errors in the steering direction and reduced robustness to reverberation.

For convenience, the SHD signal model in (6.2) can also be expressed in vector form as

$$\mathbf{p}(k) = \mathbf{x}(k) + \mathbf{v}(k) \tag{6.6}$$

where the SHD signal vector $\mathbf{p}(k)$ of length $(L+1)^2$ is defined as

$$\mathbf{p}(k) = \left[P_{00}(k)\ P_{1(-1)}(k)\ P_{10}(k)\ P_{11}(k)\ P_{2(-2)}(k)\ \cdots\ P_{LL}(k)\right]^{\mathrm{T}},$$

and $\mathbf{x}(k)$ and $\mathbf{v}(k)$ are defined similarly to $\mathbf{p}(k)$. The beamformer output signal $Z(k)$ can be expressed as

$$Z(k) = \mathbf{w}^{\mathrm{H}}(k)\mathbf{p}(k), \tag{6.7}$$

where the filter weights vector is defined as

$$\mathbf{w}(k) = \left[W_{00}(k)\ W_{1(-1)}(k)\ W_{10}(k)\ W_{11}(k)\ W_{2(-2)}(k)\ \cdots\ W_{LL}(k)\right]^{\mathrm{T}}.$$

In matrix form the desired signal is written as

$$\mathbf{x}(k, \Omega_{\mathrm{s}}) = \sqrt{P_{\mathrm{pw}}(k)}\mathbf{B}(k)\mathbf{y}^{*}(\Omega_{\mathrm{s}}), \tag{6.8}$$

where the vector of spherical harmonics $\mathbf{y}(\Omega_{\mathrm{s}})$ of length $(L+1)^2$ is defined as

$$\mathbf{y}(\Omega_{\mathrm{s}}) = \left[Y_{00}(\Omega_{\mathrm{s}})\ Y_{1(-1)}(\Omega_{\mathrm{s}})\ Y_{10}(\Omega_{\mathrm{s}})\ Y_{11}(\Omega_{\mathrm{s}})\ \cdots\ Y_{LL}(\Omega_{\mathrm{s}})\right]^{\mathrm{T}}, \tag{6.9}$$

and the $(L+1)^2 \times (L+1)^2$ matrix of mode strengths $\mathbf{B}(k)$ is defined as

$$\mathbf{B}(k) = \mathrm{diag}\left\{b_0(k), b_1(k), b_1(k), b_1(k), b_2(k), \ldots, b_L(k)\right\}, \tag{6.10}$$

therefore $\mathbf{B}(k)$ consists of $2l+1$ repetitions of $b_l(k)$ for $l \in \{0, \ldots, L\}$ along its diagonal. Finally, the distortionless constraint is given by

$$\mathbf{w}^{\mathrm{H}}(k)\mathbf{B}(k)\mathbf{y}^{*}(\Omega_{\mathrm{u}}) = 1. \tag{6.11}$$

6.2 Design Criteria

In this section, we introduce a number of measures that can be used to design optimal beamformers as in Sect. 6.3. It should be noted that these measures are defined with respect to the signals with physical significance, namely the spatial domain signals, and not with respect to the eigenbeams. Nevertheless, these measures will still depend on the eigenbeams as they form a part of the spherical harmonic expansion (SHE) of the spatial domain signals.

6.2.1 Directivity

Directivity is a measure of a beamformer's spatial selectivity and quantifies its ability to suppress sound waves that do not originate from a specifically chosen steering direction. It is defined as the ratio of the power of the beamformer output due to a plane wave arriving from the steering direction Ω_u to the power of the beamformer output averaged over all directions [28]. The directivity $\mathcal{D}(k)$ is therefore written as

$$\mathcal{D}(k) = \frac{|Z(k,\Omega_\mathrm{u})|^2}{\frac{1}{4\pi}\int_{\Omega\in\mathcal{S}^2}|Z(k,\Omega)|^2\,\mathrm{d}\Omega} \tag{6.12}$$

$$= \frac{\left|\sum_{l=0}^{L}\sum_{m=-l}^{l} W_{lm}^*(k)X_{lm}(k,\Omega_\mathrm{u})\right|^2}{\frac{1}{4\pi}\int_{\Omega\in\mathcal{S}^2}\left|\sum_{l=0}^{L}\sum_{m=-l}^{l} W_{lm}^*(k)X_{lm}(k,\Omega)\right|^2\mathrm{d}\Omega}, \tag{6.13}$$

where the notation $\int_{\Omega\in\mathcal{S}^2}\mathrm{d}\Omega$ is used to denote compactly the solid angle $\int_{\phi=0}^{2\pi}\int_{\theta=0}^{\pi}\sin\theta\mathrm{d}\theta\mathrm{d}\phi$. Applying the distortionless constraint (6.5), and by substituting the expression for a plane wave (6.4) into (6.12), we find

$$\mathcal{D}(k) = \frac{4\pi P_\mathrm{pw}(k)}{\int_{\Omega\in\mathcal{S}^2}\left|\sum_{l=0}^{L}\sum_{m=-l}^{l} W_{lm}^*(k)\sqrt{P_\mathrm{pw}(k)}b_l(k)Y_{lm}^*(\Omega)\right|^2\mathrm{d}\Omega}$$

$$= \frac{4\pi}{\int_{\Omega\in\mathcal{S}^2}\left|\sum_{l=0}^{L}\sum_{m=-l}^{l} W_{lm}^*(k)b_l(k)Y_{lm}^*(\Omega)\right|^2\mathrm{d}\Omega}. \tag{6.14}$$

Using the orthonormality of the spherical harmonics (2.18), this can be simplified to[4]

$$\mathcal{D}(k) = 4\pi\left(\sum_{l=0}^{L}\sum_{m=-l}^{l}\left|W_{lm}^*(k)b_l(k)\right|^2\right)^{-1}, \tag{6.15}$$

---

[4]It should be noted that this simplified expression is only valid for beamformers that satisfy the distortionless constraint given in (6.5). It therefore does not apply to the plane-wave decomposition beamformer presented in Sect. 6.3.1.1, which satisfies a scaled version of this constraint.

or in vector form

$$\mathcal{D}(k) = 4\pi \left|\left|\mathbf{B}(k)\mathbf{w}^*(k)\right|\right|^{-2}, \tag{6.16}$$

where $||\cdot||$ denotes the 2-norm. The directivity is therefore a function of the array properties, such as radius or microphone type, and the beamformer weights $W_{lm}(k)$.

The directivity is frequently expressed in dB and is then referred to as the *directivity index (*DI),

$$\mathrm{DI}(k) = 10 \log_{10} \mathcal{D}(k). \tag{6.17}$$

6.2.2 *Front-to-Back Ratio*

The *front-to-back ratio* is another alternative measure of a beamformer's spatial selectivity and quantifies its ability to differentiate between sound waves that originate from the front and the back. It is defined as the ratio of the average power of the beamformer output due to a plane waves arriving from the front to the average power of the beamformer output due to plane waves arriving from the back. The front-to-back ratio $\mathcal{F}(k)$ is therefore written as [7]

$$\mathcal{F}(k) = \frac{\frac{1}{4\pi}\int_{\Omega\in\mathcal{S}^2_\mathrm{F}} \left|\sum_{l=0}^{L}\sum_{m=-l}^{l} W^*_{lm}(k)X_{lm}(k,\Omega)\right|^2 \mathrm{d}\Omega}{\frac{1}{4\pi}\int_{\Omega\in\mathcal{S}^2_\mathrm{B}} \left|\sum_{l=0}^{L}\sum_{m=-l}^{l} W^*_{lm}(k)X_{lm}(k,\Omega)\right|^2 \mathrm{d}\Omega}, \tag{6.18}$$

where for a beamformer steered to $(\pi/2, \pi/2)$ we have

$$\int_{\Omega\in\mathcal{S}^2_\mathrm{F}} \mathrm{d}\Omega = \int_{\phi=0}^{\pi}\int_{\theta=0}^{\pi} \sin\theta \mathrm{d}\theta \mathrm{d}\phi \tag{6.19}$$

and

$$\int_{\Omega\in\mathcal{S}^2_\mathrm{B}} \mathrm{d}\Omega = \int_{\phi=\pi}^{2\pi}\int_{\theta=0}^{\pi} \sin\theta \mathrm{d}\theta \mathrm{d}\phi. \tag{6.20}$$

6.2.3 *White Noise Gain*

White noise gain (WNG) is a measure of a beamformer's robustness against sensor noise and errors in microphone placement and steering direction [10], and is defined as the array gain in the presence of spatially incoherent noise [28], i.e., the ratio of the signal-to-noise ratio (SNR) at the beamformer output (oSNR) to the SNR at the beamformer input (iSNR).

We now derive the WNG for a spherical microphone array employing a set of microphones uniformly distributed on the sphere. The desired signal power is different at each microphone, particularly for a rigid sphere where the scattering effects depend on the angle of incidence [16]. When calculating the iSNR, the desired signal power is therefore averaged over the sphere.

Let us assume that the noise at each microphone has equal power $\sigma_v^2(k)$. The input SNR is then given by

$$\mathrm{iSNR_w}(k) = \frac{\frac{1}{4\pi}\int_{\Omega\in\mathcal{S}^2} |X(k,\mathbf{r})|^2\,\mathrm{d}\Omega}{\sigma_v^2(k)} \tag{6.21a}$$

$$= \frac{\frac{1}{4\pi}\int_{\Omega\in\mathcal{S}^2} \left|\sum_{l=0}^{\infty}\sum_{m=-l}^{l} X_{lm}(k)Y_{lm}(\Omega)\right|^2 \mathrm{d}\Omega}{\sigma_v^2(k)}, \tag{6.21b}$$

where (6.21b) is obtained using the spherical harmonic decomposition of $X(k,\mathbf{r})$. Assuming plane-wave incidence from a direction Ω_{s}, by substituting (6.4) into (6.21), we find

$$\mathrm{iSNR_w}(k) = \frac{\int_{\Omega\in\mathcal{S}^2} \left|\sum_{l=0}^{\infty}\sum_{m=-l}^{l} \sqrt{P_{\mathrm{pw}}(k)}b_l(k)Y_{lm}^*(\Omega_{\mathrm{s}})Y_{lm}(\Omega)\right|^2 \mathrm{d}\Omega}{4\pi\sigma_v^2(k)}. \tag{6.22}$$

Using Unsöld's theorem [29], a special case of the spherical harmonic addition theorem (2.23), and the orthonormality of the spherical harmonics, we simplify (6.22) to

$$\mathrm{iSNR_w}(k) = \frac{\sum_{l=0}^{\infty}\sum_{m=-l}^{l} \left|\sqrt{P_{\mathrm{pw}}(k)}b_l(k)Y_{lm}^*(\Omega_{\mathrm{s}})\right|^2}{4\pi\sigma_v^2(k)} \tag{6.23a}$$

$$= \frac{P_{\mathrm{pw}}(k)\sum_{l=0}^{\infty} |b_l(k)|^2\,(2l+1)}{(4\pi)^2\sigma_v^2(k)}. \tag{6.23b}$$

The input SNR is therefore a function of the plane wave power $P_{\mathrm{pw}}(k)$, the array properties, via the mode strength $b_l(k)$, and the noise power $\sigma_v^2(k)$.

The output SNR is given by

$$\mathrm{oSNR_w}(k) = \frac{\left|\sum_{l=0}^{L}\sum_{m=-l}^{l} W_{lm}^*(k)X_{lm}(k)\right|^2}{\mathrm{E}\left\{\left|\sum_{l=0}^{L}\sum_{m=-l}^{l} W_{lm}^*(k)V_{lm}(k)\right|^2\right\}}. \tag{6.24}$$

Applying the distortionless constraint (6.5), this reduces to

$$\mathrm{oSNR_w}(k) = \frac{P_{\mathrm{pw}}(k)}{\mathrm{E}\left\{\left|\sum_{l=0}^{L}\sum_{m=-l}^{l} W_{lm}^*(k) V_{lm}(k)\right|^2\right\}}. \tag{6.25}$$

With Q microphones uniformly distributed on the sphere, the cross power spectral density of the noise is given by [31, Eq. 7.31]

$$\mathrm{E}\left\{V_{lm}(k)V_{l'm'}^*(k)\right\} = \sigma_v^2(k)\frac{4\pi}{Q}\delta_{l,l'}\delta_{m,m'}, \tag{6.26}$$

where δ denotes the Kronecker delta, and oSNR simplifies to

$$\mathrm{oSNR_w}(k) = P_{\mathrm{pw}}(k)\left(\frac{4\pi}{Q}\sigma_v^2(k)\sum_{l=0}^{L}\sum_{m=-l}^{l}\left|W_{lm}^*(k)\right|^2\right)^{-1}. \tag{6.27}$$

The output SNR is a function of the beamformer weights $W_{lm}(k)$, the plane wave power $P_{\mathrm{pw}}(k)$, the noise power $\sigma_v^2(k)$, and the beamformer order L. The beamformer order can be increased by adding microphones, as discussed in Sect. 3.4.

Finally, the WNG can be expressed as

$$\mathrm{WNG}(k) = \frac{\mathrm{oSNR_w}(k)}{\mathrm{iSNR_w}(k)} \tag{6.28a}$$

$$= \frac{4\pi Q}{||\mathbf{w}(k)||^2 \sum_{l=0}^{\infty} |b_l(k)|^2 (2l+1)}. \tag{6.28b}$$

The WNG is a function of the beamformer weights $W_{lm}(k)$, array order L and the array properties. As expected, it is also an increasing function of the number of microphones Q. In the case of an open sphere, $b_l(k) = i^l j_l(kr)$, and since $\sum_{l=0}^{\infty} |j_l(kr)|^2 (2l+1) = 1$ [2, 13], the WNG is given by the simple expression

$$\mathrm{WNG}(k) = \frac{4\pi Q}{||\mathbf{w}(k)||^2}. \tag{6.29}$$

6.2.4 Spatial Response

The output of the beamformer in the presence of a single unit amplitude plane wave originating from a DOA Ω is given by

$$\mathcal{B}(k, \Omega) = \mathbf{w}^{\mathrm{H}}(k)\mathbf{B}(k)\mathbf{y}^*(\Omega), \tag{6.30}$$

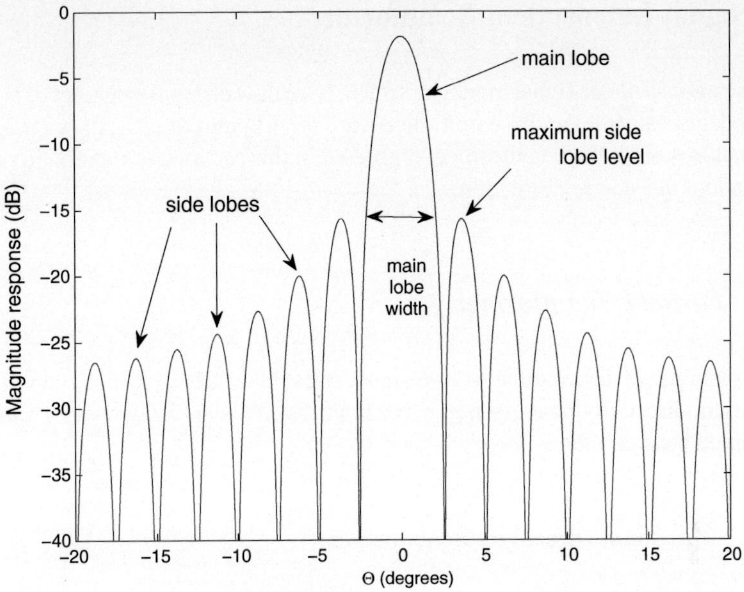

Fig. 6.2 Illustrative example of the magnitude of a spatial response $\mathcal{B}(k, \Theta)$ as a function of the angle Θ between the steering direction and DOA

and is known as the *spatial response* of the beamformer. The square magnitude of the spatial response $\mathcal{B}(k, \Omega)$ is referred to as the *beam pattern* [4].[5] The beam pattern describes the beamformer's ability to select signals originating from a direction of interest, while suppressing signals that do not. Beam patterns typically exhibit multiple peaks or *lobes*; the largest lobe, in the direction of interest, is referred to as the *main lobe*, while the other lobes are referred to as *sidelobes*. Due to the effects of spatial aliasing, some sidelobes may have an amplitude equal to that of the main lobe, and they are then referred to as *grating lobes* [27].

Due to the spherical symmetry of the SHD, the beam pattern can also be expressed as a function of the angle between the DOA Ω and the beamformer's steering direction Ω_{u}, denoted as Θ. Ideally, the response in the steering direction, $\mathcal{B}(k, \Theta = 0)$, should be as large as possible compared to the response in other directions, i.e., the *sidelobe levels* should be minimized. We refer to the width of the region that has a higher response than the maximum sidelobe level as the *main lobe width*,[6] as illustrated in Fig. 6.2.

[5]Note that in some publications, such as [28], $\mathcal{B}(k, \Omega)$ is referred to as the beam pattern, and its square magnitude is referred to as the *power pattern*.

[6]The main lobe width is sometimes also defined as the width of the region where the beam pattern is no less than half of its maximum value, or equivalently, no more than 3 dB below its maximum value.

6.3 Signal-Independent Beamformers

Having established our signal model in Sect. 6.1, we now develop a number of signal-independent beamformers based on the design criteria introduced in Sect. 6.2. The beam patterns of all the beamformers presented in this section are rotationally symmetric about the steering direction.

6.3.1 *Farfield Beamformers*

In this section, we derive three beamformers suitable for use in farfield conditions: a maximum directivity beamformer, a maximum WNG beamformer, and a multiply constrained beamformer.

6.3.1.1 Maximum Directivity Beamformer

The beamformer that maximizes the directivity while imposing a distortionless constraint in the steering direction satisfies

$$\max_{\mathbf{w}(k)} \mathcal{D}(k) \quad \text{subject to} \quad \mathbf{w}^{\mathrm{H}}(k)\mathbf{B}(k)\mathbf{y}^*(\Omega_{\mathrm{u}}) = 1,$$

or equivalently,

$$\min_{\mathbf{w}(k)} \left|\left|\mathbf{B}(k)\mathbf{w}^*(k)\right|\right|^2 \quad \text{subject to} \quad \mathbf{w}^{\mathrm{H}}(k)\mathbf{B}(k)\mathbf{y}^*(\Omega_{\mathrm{u}}) = 1,$$

where $\mathbf{y}(\Omega_{\mathrm{u}})$ is the vector of spherical harmonics defined in (6.9).

Following the approach proposed by Brandwood [5], if we use a Lagrange multiplier to adjoin the constraint to the cost function, the weights of the maximum directivity beamformer are then given by

$$\mathbf{w}_{\mathrm{maxDI}}(k) = \arg\min_{\mathbf{w}(k)} \mathcal{L}(\mathbf{w}(k), \lambda), \tag{6.31}$$

where $\mathcal{L}$ is the complex Lagrangian given by

$$\begin{aligned}\mathcal{L}(\mathbf{w}(k), \lambda) = &\left[\mathbf{B}(k)\mathbf{w}^*(k)\right]^{\mathrm{H}} \left[\mathbf{B}(k)\mathbf{w}^*(k)\right] \\ &+ \lambda \left(\mathbf{w}^{\mathrm{H}}(k)\mathbf{B}(k)\mathbf{y}^*(\Omega_{\mathrm{u}}) - 1\right) + \lambda^* \left(\mathbf{y}^{\mathrm{T}}(\Omega_{\mathrm{u}})\mathbf{B}^*(k)\mathbf{w}(k) - 1\right)\end{aligned} \tag{6.32}$$

and λ is the Lagrange multiplier. Setting the gradient of $\mathcal{L}(\mathbf{w}_{\mathrm{maxDI}}(k), \lambda)$ with respect to $\mathbf{w}^*_{\mathrm{maxDI}}$ to zero yields

$$\begin{aligned}\nabla_{\mathbf{w}^*_{\mathrm{maxDI}}} \mathcal{L}(\mathbf{w}_{\mathrm{maxDI}}(k), \lambda) &= \mathbf{0}_N \\ \mathbf{B}(k)\mathbf{B}^*(k)\mathbf{w}_{\mathrm{maxDI}}(k) + \lambda \mathbf{B}(k)\mathbf{y}^*(\Omega_{\mathrm{u}}) &= \mathbf{0}_N,\end{aligned} \tag{6.33}$$

where $\mathbf{0}_N$ is a column vector of N zeros. Using the constraint in (6.31), we then find

$$\mathbf{w}_{\mathrm{maxDI}}(k) = \frac{[\mathbf{B}^*(k)]^{-1}\,\mathbf{y}^*(\Omega_{\mathrm{u}})}{||\mathbf{y}(\Omega_{\mathrm{u}})||^2}. \tag{6.34}$$

Using Unsöld's theorem [29], this simplifies to

$$\mathbf{w}_{\mathrm{maxDI}}(k) = \frac{4\pi}{(L+1)^2}\left[\mathbf{B}^*(k)\right]^{-1}\mathbf{y}^*(\Omega_{\mathrm{u}}), \tag{6.35}$$

or in scalar form

$$W_{lm}^{\mathrm{maxDI}}(k) = \frac{4\pi}{(L+1)^2}\frac{Y_{lm}^*(\Omega_{\mathrm{u}})}{b_l^*(k)}. \tag{6.36}$$

A well-known farfield SHD beamformer is the *plane-wave decomposition* (PWD) beamformer, also sometimes known as a *regular* beamformer [24], whose weights are given by [22]

$$\mathbf{w}_{\mathrm{PWD}}(k) = \left[\mathbf{B}^*(k)\right]^{-1}\mathbf{y}^*(\Omega_{\mathrm{u}}). \tag{6.37}$$

As the (frequency-independent) scaling factor does not affect the directivity, the PWD beamformer is also a maximum directivity beamformer. The reason for the name PWD will become clear in the next paragraph.

Assuming a single unit amplitude plane wave is incident upon the array from a direction Ω_{s}, the output $Z(k)$ of the PWD beamformer is given by

$$Z(k) = \mathbf{w}^{\mathrm{H}}(k)\mathbf{B}(k)\mathbf{y}^*(\Omega_{\mathrm{s}}) \tag{6.38a}$$

$$= \mathbf{y}^{\mathrm{T}}(\Omega_{\mathrm{u}})\mathbf{B}^{-1}(k)\mathbf{B}(k)\mathbf{y}^*(\Omega_{\mathrm{s}}) \tag{6.38b}$$

$$= \sum_{l=0}^{L}\sum_{m=-l}^{l} Y_{lm}(\Omega_{\mathrm{u}})Y_{lm}^*(\Omega_{\mathrm{s}}) \tag{6.38c}$$

$$= \begin{cases} \dfrac{(L+1)^2}{4\pi} & \text{if } \Omega_{\mathrm{s}} = \Omega_{\mathrm{u}}, \\ \dfrac{(L+1)}{4\pi(\cos\Theta - 1)}\left[\mathcal{P}_{L+1}(\cos\Theta) - \mathcal{P}_L(\cos\Theta)\right] & \text{otherwise,} \end{cases} \tag{6.38d}$$

where Θ is the angle between Ω_{s} and Ω_{u} and $\mathcal{P}_L$ is the Legendre polynomial of order L. The Christoffel summation formula [11, Sect. 8.915] is used to obtain (6.38d) [20]. The beamformer output $Z(k)$ reaches its maximum when $\Theta = 0$, such that the steering direction Ω_{u} is equal to the arrival direction Ω_{s}, as desired. We normalize the

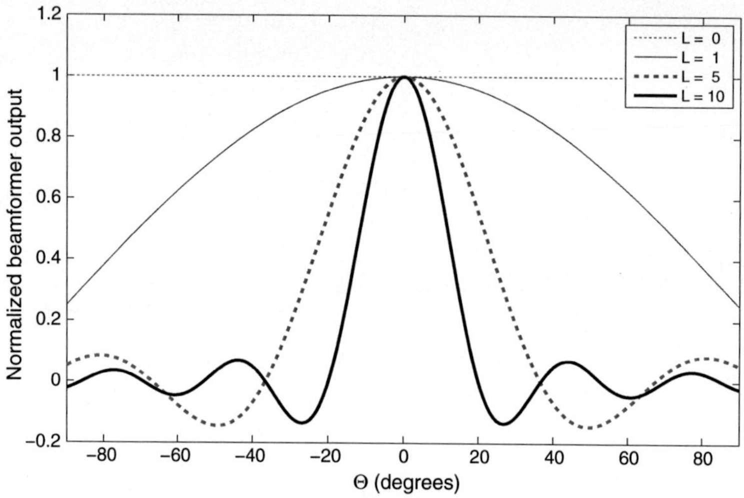

Fig. 6.3 Normalized beamformer output as a function of the beamformer order L and Θ, the angle between the beamformer steering direction and the DOA

beamformer output with respect to its value for $\Theta = 0$, and plot it as a function of Θ in Fig. 6.3. We see that as L increases, the distribution of $Z(k)$ narrows around $\Theta = 0$, tending towards a delta function for $L \to \infty$ [31, Eq. 6.47].

The directivity of the maximum directivity beamformer is given by substituting (6.35) into (6.16)[7]

$$\mathcal{D}(k) = 4\pi \left|\left| \frac{4\pi}{(L+1)^2} \mathbf{B}(k)\mathbf{B}^{-1}(k)\mathbf{y}(\Omega_\mathrm{u}) \right|\right|^{-2} \tag{6.39a}$$

$$= \frac{(L+1)^4}{4\pi} \, ||\mathbf{y}(\Omega_\mathrm{u})||^{-2} \tag{6.39b}$$

$$= (L+1)^2. \tag{6.39c}$$

The directivity of the maximum directivity beamformer is therefore frequency-independent and only depends on the beamformer order L.

Since at least $(L+1)^2$ microphones are required to sample a sound field up to order L without spatial aliasing, the directivity is upper bounded by the number of microphones Q. This is also the maximum directivity of a spatial domain beamformer based on a standard linear array [28, Eq. 2.160].

The WNG of the maximum directivity beamformer is given by substituting (6.35) into (6.28)

[7]This expression is identical to (12) in [22] if we substitute $d_n = 1$.

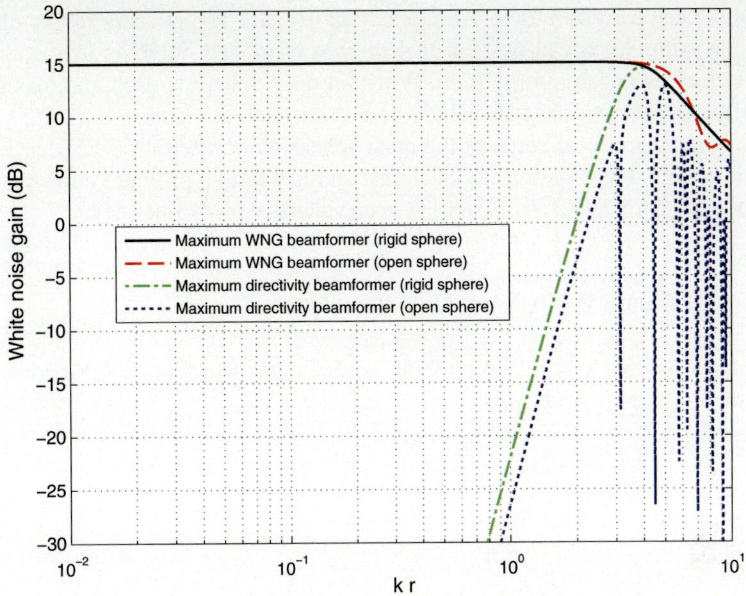

Fig. 6.4 WNG of the maximum directivity and maximum WNG beamformers of order $L = 4$ as a function of kr, for open and rigid arrays

$$\mathrm{WNG}(k) = \frac{Q(L+1)^4}{4\pi \sum_{l=0}^{L} \sum_{m=-l}^{l} \left| \frac{Y_{lm}(\Omega_{\mathrm{u}})}{b_l(k)} \right|^2 \sum_{l=0}^{\infty} |b_l(k)|^2 (2l+1)} \tag{6.40a}$$

$$= \frac{Q(L+1)^4}{\sum_{l=0}^{L} |b_l(k)|^{-2} (2l+1) \sum_{l=0}^{\infty} |b_l(k)|^2 (2l+1)}. \tag{6.40b}$$

In the open sphere case, this simplifies to[8]

$$\mathrm{WNG}(k) = \frac{Q(L+1)^4}{\sum_{l=0}^{L} |b_l(k)|^{-2} (2l+1)}, \tag{6.41}$$

or in matrix form

$$\mathrm{WNG}(k) = Q(L+1)^4 \left|\left| \mathbf{B}^{-1}(k) \right|\right|^{-2}. \tag{6.42}$$

In Fig. 6.4, we plot the WNG of the maximum directivity beamformer of order $L = 4$ as a function of the product of the wavenumber k and array radius r, kr, for an array of $Q = 32$ microphones. Assuming a speed of sound of $343\,\mathrm{m} \cdot \mathrm{s}^{-1}$,

[8]This expression is identical to (11) in [22] if we substitute $d_n = 1$, with the exception of the $(4\pi)^2$ scaling factor, which is required due to the fact that in [22] a 4π scaling factor is included in the definition of the mode strength.

a kr value of 1 corresponds to a frequency of 1.1 kHz for an array radius of $r =$ 10 cm, for example. It can be seen that the beamformer's WNG is low except at high frequencies or large array radii. When an open sphere is used, the maximum directivity beamformer has particularly poor robustness at certain values of kr; this is due to the presence of zeros in the open sphere mode strength (see Sect. 3.4.2). The rigid sphere does not present this issue, and in addition provides an increase in WNG of approximately 3.7 dB over the open sphere at low values of kr.

6.3.1.2 Maximum White Noise Gain Beamformer

The beamformer that maximizes the WNG while imposing a distortionless constraint in the steering direction satisfies

$$\max_{\mathbf{w}(k)} \text{WNG}(k) \quad \text{subject to} \quad \mathbf{w}^{\mathrm{H}}(k)\mathbf{B}(k)\mathbf{y}^*(\Omega_{\mathrm{u}}) = 1,$$

or equivalently,

$$\min_{\mathbf{w}(k)} ||\mathbf{w}(k)||^2 \quad \text{subject to} \quad \mathbf{w}^{\mathrm{H}}(k)\mathbf{B}(k)\mathbf{y}^*(\Omega_{\mathrm{u}}) = 1.$$

Proceeding in a similar way as for the analysis of the maximum directivity beamformer, if we use a Lagrange multiplier to adjoin the constraint to the cost function, the weights of the maximum directivity beamformer are then given by

$$\mathbf{w}_{\mathrm{maxWNG}}(k) = \arg\min_{\mathbf{w}(k)} \mathcal{L}(\mathbf{w}(k), \lambda), \tag{6.43}$$

where $\mathcal{L}$ is the complex Lagrangian given by

$$\begin{aligned}\mathcal{L}(\mathbf{w}(k), \lambda) = {} & [\mathbf{w}(k)]^{\mathrm{H}} [\mathbf{w}(k)] \\ & + \lambda \left(\mathbf{w}^{\mathrm{H}}(k)\mathbf{B}(k)\mathbf{y}^*(\Omega_{\mathrm{u}}) - 1\right) + \lambda^* \left(\mathbf{y}^{\mathrm{T}}(\Omega_{\mathrm{u}})\mathbf{B}^*(k)\mathbf{w}(k) - 1\right)\end{aligned} \tag{6.44}$$

and λ is the Lagrange multiplier. Setting the gradient of $\mathcal{L}(\mathbf{w}_{\mathrm{maxWNG}}(k), \lambda)$ with respect to $\mathbf{w}^*_{\mathrm{maxWNG}}$ to zero yields

$$\begin{aligned}\nabla_{\mathbf{w}^*_{\mathrm{maxWNG}}} \mathcal{L}(\mathbf{w}_{\mathrm{maxWNG}}(k), \lambda) &= \mathbf{0}_N \\ \mathbf{w}_{\mathrm{maxWNG}}(k) + \lambda \mathbf{B}(k)\mathbf{y}^*(\Omega_{\mathrm{u}}) &= \mathbf{0}_N,\end{aligned} \tag{6.45}$$

where $\mathbf{0}_N$ is a column vector of N zeros. Using the constraint in (6.43), we then find

$$\mathbf{w}_{\mathrm{maxWNG}}(k) = \frac{\mathbf{B}(k)\mathbf{y}^*(\Omega_{\mathrm{u}})}{\mathbf{y}^{\mathrm{T}}(\Omega_{\mathrm{u}})\mathbf{B}^*(k)\mathbf{B}(k)\mathbf{y}^*(\Omega_{\mathrm{u}})}. \tag{6.46}$$

Using Unsöld's theorem [29], this simplifies to

$$\mathbf{w}_{\text{maxWNG}}(k) = 4\pi \frac{\mathbf{B}(k)\mathbf{y}^*(\Omega_\text{u})}{||\mathbf{B}(k)||^2}, \tag{6.47}$$

or in scalar form

$$W_{lm}^{\text{maxWNG}}(k) = 4\pi \frac{Y_{lm}^*(\Omega_\text{u}) b_l(k)}{\sum_{l=0}^{L} |b_l(k)|^2 (2l+1)}. \tag{6.48}$$

A well-known farfield SHD beamformer is the *delay-and-sum* beamformer, whose weights are given by [22]

$$\mathbf{w}_{\text{DSB}}(k) = \mathbf{B}(k)\mathbf{y}^*(\Omega_\text{u}). \tag{6.49}$$

In the case of an open sphere, $b_l(k) = i^l j_l(kr)$, and since $\sum_{l=0}^{\infty} |j_l(kr)|^2 (2l+1) = 1$ [2], the following relationship between the maximum WNG and delay-and-sum beamformers is obtained:

$$\lim_{L\to\infty} \mathbf{w}_{\text{maxWNG}}(k) = 4\pi\, \mathbf{w}_{\text{DSB}}(k). \tag{6.50}$$

When an open sphere is used, the delay-and-sum beamformer therefore approaches a maximum WNG beamformer as $L \to \infty$ (ignoring the 4π scaling factor, which does not affect the WNG). For a finite L and/or if another microphone type or array configuration is used (such as a rigid sphere), the delay-and-sum beamformer is slightly suboptimal.

The delay-and-sum beamformer owes its name to the fact that for an open sphere as $L \to \infty$, its output converges to the output of the widely known spatial domain delay-and-sum beamformer [22].

The directivity of the maximum WNG beamformer is given by substituting (6.47) into (6.16)

$$\mathcal{D}(k) = 4\pi \left|\left| \frac{4\pi}{||\mathbf{B}(k)||^2} \mathbf{B}(k)\mathbf{B}^*(k)\mathbf{y}(\Omega_\text{u}) \right|\right|^{-2} \tag{6.51a}$$

$$= \frac{4\pi}{(4\pi)^2} ||\mathbf{B}(k)||^4 \left|\left|\mathbf{B}(k)\mathbf{B}^*(k)\mathbf{y}(\Omega_\text{u})\right|\right|^{-2} \tag{6.51b}$$

$$= ||\mathbf{B}(k)||^4 \left|\left|\mathbf{B}(k)\mathbf{B}^*(k)\right|\right|^{-2}. \tag{6.51c}$$

The WNG of the maximum WNG beamformer is given by substituting (6.47) into (6.28)

$$\text{WNG}(k) = \frac{4\pi Q\, ||\mathbf{B}(k)||^4}{(4\pi)^2\, ||\mathbf{B}(k)\mathbf{y}^*(\Omega_\text{u})||^2 \sum_{l=0}^{\infty} |b_l(k)|^2 (2l+1)} \tag{6.52a}$$

Using Unsöld's theorem [29], this simplifies to

$$\mathrm{WNG}(k) = \frac{Q\,||\mathbf{B}(k)||^2}{\sum_{l=0}^{\infty} |b_l(k)|^2\,(2l+1)} \tag{6.53}$$

In the open sphere case, the WNG approaches Q as $L \to \infty$ (as in [22]), so it can be seen that the maximum WNG beamformer achieves a constant WNG of Q that is independent of frequency. This is also the highest achievable WNG for a distortionless beamformer in the spatial domain [28].

In Fig. 6.5, we plot the DI of the maximum directivity and maximum WNG beamformers of order $L = 4$ as a function of kr for an array of $Q = 32$ microphones. As expected, the maximum directivity beamformer provides the highest directivity; while the maximum WNG beamformer has poor directivity at low values of kr (i.e., low frequencies or small array radii). Due to the effects of scattering introduced by the rigid sphere (see Sect. 3.4.1), the maximum WNG beamformer has better directivity with a rigid array than with an open array. The directivity of the maximum directivity beamformer is independent of kr, while for the maximum WNG beamformer the directivity decays as kr decreases, tending towards 0 dB (i.e., no directivity).

The WNG of the maximum WNG beamformer of order $L = 4$ is shown in Fig. 6.4; as expected, it provides the highest WNG. Using Figs. 6.4 and 6.5, it can be observed that there is a tradeoff between WNG and directivity. The maximum directivity and

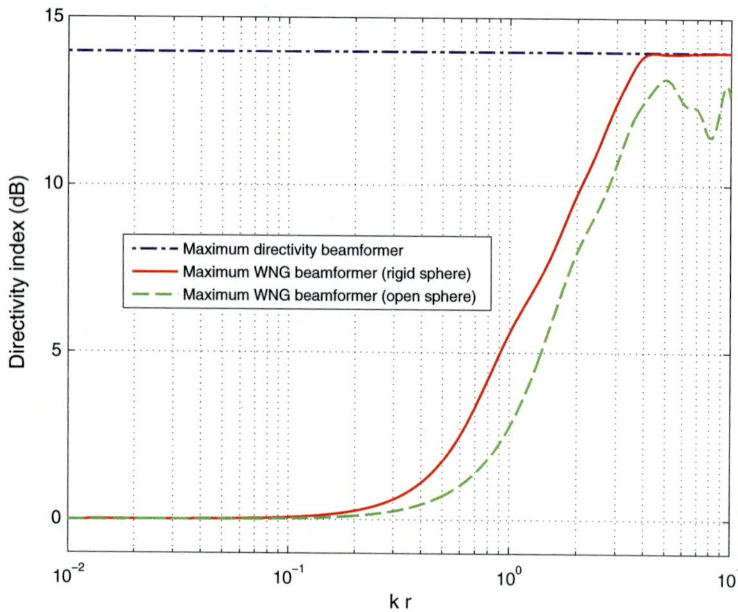

Fig. 6.5 Directivity of the maximum directivity and maximum WNG beamformers of order $L = 4$ as a function of kr

WNG beamformers provide performance bounds for SHD beamformers in terms of directivity and WNG, and are attractive due to their low computational complexity. However, in practice a compromise solution is desirable, such as the multiply constrained beamformer presented in Sect. 6.3.1.3, or the signal-dependent beamformers in Chap. 7, which adaptively control the tradeoff between these two objectives depending on the nature of the noise to be suppressed.

6.3.1.3 Multiply Constrained Beamformer

Another approach to the design of a signal-independent beamformer is to minimize its sidelobe levels for a given main lobe width, to ensure that interfering signals that do not originate from the steering direction are effectively suppressed. However, in order to obtain a beamformer that is robust to errors in sensor position and steering direction, and to sensor noise, it is desirable to introduce a constraint on the beamformer's WNG.

In [25], the authors propose a robust minimum sidelobe beamformer, which minimizes the maximum sidelobe level, subject to a distortionless constraint in the steering direction and a minimum WNG constraint. The objective can therefore be expressed in the form of a *minimax criterion* as

$$\begin{aligned} &\min_{\mathbf{w}(k)} \quad \max_{\Theta > \Delta/2} |\mathcal{B}(k, \Theta)| \quad \text{subject to} \\ &\quad \mathbf{w}^{\mathrm{H}}(k)\mathbf{B}(k)\mathbf{y}^{*}(\Omega_{\mathrm{u}}) = 1, \quad \mathrm{WNG}(k) \geq \zeta(k), \end{aligned} \tag{6.54}$$

where $\mathcal{B}(k, \Theta)$ is the spatial response of the beamformer, Θ denotes the angle between the steering direction and the DOA, Δ denotes the main lobe width (as defined in Sect. 6.2.4), and ζ is the minimum WNG. The sidelobe region is defined as $\Theta_{\mathrm{SL}} = \{\Theta | \Theta > \Delta/2\}$.

As shown in [25], the problem in (6.54) can be reformulated as a convex optimization problem, solvable using second-order cone programming. The sidelobe region is approximated using a finite grid $\Theta_{n_{\mathrm{g}}} \in \Theta_{\mathrm{SL}}, n_{\mathrm{g}} \in \{1, \ldots, N_{\mathrm{g}}\}$; the approximation then improves as N_{g} increases.

Finding a solution to (6.54) can be computationally intensive. However, a significant advantage of SHD beamforming is that if the desired beam pattern is rotationally symmetric about the steering direction Ω_{u}, the process of computing the beamformer weights and steering of the beamformer can be decoupled. In this case, the beamformer weights are expressed as $W_{lm}(k) = C_l(k) Y_{lm}^{*}(\Omega_{\mathrm{u}})$, and the weights $C_l(k)$ then become the quantities to be optimized. If the desired beam pattern is not rotationally symmetric about the steering direction, the beam pattern can be rotated by multiplying the SHD beamformer weights by Wigner-D functions that depend on the rotation angles, as proposed in [23].

6.3.2 Nearfield Beamformers

In this chapter, we have until now assumed that the desired signal was due to a single plane wave, i.e., farfield conditions. However, under nearfield conditions, the plane wave assumptions cannot be considered valid. The SHD sound pressure due to a spherical wave originating from a source at a position $\mathbf{r}_s = (r_s, \Omega_s)$ is given by

$$X_{lm}(k, \mathbf{r}_s) = X_{sw}(k) b_l^{nf}(k, r_s) Y_{lm}^*(\Omega_s), \tag{6.55}$$

where $X_{sw}(k)$ denotes the spherical wave amplitude and the nearfield mode strength $b_l^{nf}(k, r_s)$ is given by

$$b_l^{nf}(k, r_s) = -iki^{-l} h_l^{(2)}(kr_s) b_l(k), \tag{6.56}$$

and $h_l^{(2)}$ is the spherical Hankel function of the second kind and of order l.

Beamformers suitable for nearfield conditions [8, 9, 19] can be designed by replacing the farfield mode strength expression $b_l(k)$ with the nearfield mode strength $b_l^{nf}(k, r_s)$ in the beamformer weights. For example, the weights of a nearfield plane-wave decomposition beamformer are given by

$$W_{lm}^{PWD,nf}(k) = \frac{Y_{lm}^*(\Omega_u)}{\left[b_l^{nf}(k, r_s)\right]^*}, \tag{6.57}$$

instead of (6.37). While this process is straightforward, it does require knowledge of the source-array distance r_s. If the source-array knowledge is not known, the source-array distance r_s becomes a controllable parameter, which is effectively a look distance and enables radial discrimination [9].

An appropriate boundary between the farfield and nearfield regions can be determined by comparing the magnitudes of the farfield mode strength $b_l(k)$ and the nearfield mode strength $b_l^{nf}(k, r_s)$, as proposed in [8]. Using this criterion, the cut-off distance r_{nf} is determined as

$$r_{nf}(k) = \frac{L}{k}. \tag{6.58}$$

The extent of the nearfield region therefore decreases with frequency. An array with good radial discrimination, i.e., a large nearfield region, can be realized either at low frequencies (small k), or by oversampling the array (large N) [9].

Example: At a frequency of 100 Hz, assuming a speed of sound of $343\,\text{m} \cdot \text{s}^{-1}$ and an array order $L = 4$, the cut-off distance is $r_{nf}(k) = 2.2\,\text{m}$, while at a frequency of 4 kHz it is 5.5 cm.

6.4 Chapter Summary

An overview of beamforming in the SHD using signal-independent beamformers has been presented. We introduced a number of performance measures, which were then used to derive beamformers weights that are optimal with respect to these measures. We also showed the relationship between these optimal beamformers and two well-known SHD beamformers: the PWD and delay-and-sum beamformers. Finally, where similarities existed, the performance bounds for SHD beamformers were related to previously derived bounds for spatial domain beamformers.

References

1. Abhayapala, T.D., Ward, D.B.: Theory and design of high order sound field microphones using spherical microphone array. In: Proceedings of the IEEE International Conference on Acoustics, Speech and Signal Processing (ICASSP), vol. 2, pp. 1949–1952 (2002). doi:10.1109/ICASSP.2002.1006151
2. Abramowitz, M., Stegun, I.A. (eds.): Handbook of Mathematical Functions with Formulas, Graphs, and Mathematical Tables. Dover Publications, New York (1972)
3. Benesty, J., Chen, J., Huang, Y.: Microphone Array Signal Processing. Springer, Berlin (2008)
4. Brandstein, M.S., Ward, D.B. (eds.): Microphone Arrays: Signal Processing Techniques and Applications. Springer, Berlin (2001)
5. Brandwood, D.H.: A complex gradient operator and its application in adaptive array theory. Proc. IEEE **130**(1, Parts F and H), 11–16 (1983)
6. Doclo, S.: Multi-microphone noise reduction and dereverberation techniques for speech applications. Ph.D. thesis, Katholieke Universiteit Leuven, Belgium (2003)
7. Elko, G.W.: Differential microphone arrays. In: Huang, Y., Benesty, J. (eds.) Audio Signal Processing for Next-Generation Multimedia Communication Systems, pp. 2–65. Kluwer (2004)
8. Fisher, E., Rafaely, B.: The nearfield spherical microphone array. In: Proceedings of the IEEE International Conference on Acoustics, Speech and Signal Processing (ICASSP), pp. 5272–5275 (2008). doi:10.1109/ICASSP.2008.4518849
9. Fisher, E., Rafaely, B.: Near-field spherical microphone array processing with radial filtering. IEEE Trans. Audio, Speech, Lang. Process. **19**(2), 256–265 (2011). doi:10.1109/TASL.2010.2047421
10. Gilbert, E., Morgan, S.: Optimum design of directive antenna arrays subject to random variations. Bell Syst. Tech. J. **34**, 637–663 (1955)
11. Gradshteyn, I.S., Ryzhik, I.M.: Table of Integrals, Series, and Products, seventh edn. Academic Press, Cambridge (2007)
12. Habets, E.A.P., Benesty, J., Cohen, I., Gannot, S., Dmochowski, J.: New insights into the MVDR beamformer in room acoustics. IEEE Trans. Audio, Speech, Lang. Process. **18**, 158–170 (2010)
13. Jarrett, D.P., Habets, E.A.P.: On the noise reduction performance of a spherical harmonic domain tradeoff beamformer. IEEE Signal Process. Lett. **19**(11), 773–776 (2012)
14. Jarrett, D.P., Habets, E.A.P., Benesty, J., Naylor, P.A.: A tradeoff beamformer for noise reduction in the spherical harmonic domain. In: Proceedings of the International Workshop Acoustics Signal Enhancement (IWAENC). Aachen, Germany (2012)
15. Jarrett, D.P., Habets, E.A.P., Naylor, P.A.: Spherical harmonic domain noise reduction using an MVDR beamformer and DOA-based second-order statistics estimation. In: Proceedings of the IEEE International Conference on Acoustics, Speech and Signal Processing (ICASSP), pp. 654–658. Vancouver, Canada (2013)

16. Jarrett, D.P., Thiergart, O., Habets, E.A.P., Naylor, P.A.: Coherence-based diffuseness estimation in the spherical harmonic domain. In: Proceedings of the IEEE Convention of Electrical and Electronics Engineers in Israel (IEEEI). Eilat, Israel (2012)
17. Meyer, J., Agnello, T.: Spherical microphone array for spatial sound recording. In: Proceedings of the Audio Engineering Society Convention, pp. 1–9. New York (2003)
18. Meyer, J., Elko, G.: A highly scalable spherical microphone array based on an orthonormal decomposition of the soundfield. In: Proceedings of the IEEE International Conference on Acoustics, Speech and Signal Processing (ICASSP), vol. 2, pp. 1781–1784 (2002)
19. Meyer, J., Elko, G.W.: Position independent close-talking microphone. Signal Processing **86**(6), 1254–1259 (2006). doi:10.1016/j.sigpro.2005.05.036
20. Rafaely, B.: Plane-wave decomposition of the pressure on a sphere by spherical convolution. J. Acoust. Soc. Am. **116**(4), 2149–2157 (2004)
21. Rafaely, B.: Analysis and design of spherical microphone arrays. IEEE Trans. Speech Audio Process. **13**(1), 135–143 (2005). doi:10.1109/TSA.2004.839244
22. Rafaely, B.: Phase-mode versus delay-and-sum spherical microphone array processing. IEEE Signal Process. Lett. **12**(10), 713–716 (2005). doi:10.1109/LSP.2005.855542
23. Rafaely, B., Kleider, M.: Spherical microphone array beam steering using Wigner-D weighting. IEEE Signal Process. Lett. **15**, 417–420 (2008). doi:10.1109/LSP.2008.922288
24. Rafaely, B., Peled, Y., Agmon, M., Khaykin, D., Fisher, E.: Spherical microphone array beamforming. In: Cohen, I., Benesty, J., Gannot, S. (eds.) Speech Processing in Modern Communication: Challenges and Perspectives, Chap. 11. Springer, Heidelberg (2010)
25. Sun, H., Yan, S., Svensson, U.P.: Robust minimum sidelobe beamforming for spherical microphone arrays. IEEE Trans. Audio, Speech, Lang. Process. **19**(4), 1045–1051 (2011). doi:10.1109/TASL.2010.2076393
26. Teutsch, H.: Wavefield decomposition using microphone arrays and its application to acoustic scene analysis. Ph.D. thesis, Friedrich-Alexander Universität Erlangen-Nürnberg (2005)
27. van Trees, H.L.: Detection, Estimation, and Modulation Theory, Optimum Array Processing, vol. IV. Wiley, New York (2002)
28. van Trees, H.L.: Optimum Array Processing. Detection Estimation and Modulation Theory. Wiley, New York (2002)
29. Unsöld, A.: Beiträge zur Quantenmechanik der Atome. Annalen der Physik **387**(3), 355–393 (1927). doi:10.1002/andp.19273870304
30. van Veen, B.D., Buckley, K.M.: Beamforming: a versatile approach to spatial filtering. IEEE Acoust. Speech Signal Mag. **5**(2), 4–24 (1988)
31. Williams, E.G.: Fourier Acoustics: Sound Radiation and Nearfield Acoustical Holography, 1st edn. Academic Press, London (1999)

Chapter 7
Signal-Dependent Array Processing

In the previous chapter, we presented a number of signal-independent beamformers or filters,[1] including maximum directivity and maximum white noise gain beamformers. However, in practice it is desirable to control the tradeoff between these two performance objectives. In this chapter, we derive a number of signal-dependent beamformers, which adaptively seek to achieve optimal performance in terms of noise reduction and speech distortion, taking into account the statistics of the desired signal and the noise.

7.1 Signal Model

We consider a conventional frequency domain signal model in which a spherical microphone array captures a received source signal X originating from a source $\mathcal{S}$, and a noise signal V. The spatial domain signal received at Q microphone positions $\mathbf{r}_q = (r, \Omega_q) = (r, \theta_q, \phi_q), q \in \{1, \ldots, Q\}$ (in spherical coordinates, where θ_q denotes the inclination and ϕ_q denotes the azimuth) for a wavenumber k can then be expressed as[2]

$$P(k, \mathbf{r}_q) = H(k, \mathbf{r}_q)S(k) + V(k, \mathbf{r}_q) \tag{7.1a}$$

$$= X(k, \mathbf{r}_q) + V(k, \mathbf{r}_q), \tag{7.1b}$$

where $H(k, \mathbf{r}_q)$ is the acoustic transfer function from the source $\mathcal{S}$ to the microphone at position $\mathbf{r}_q$, and $S(k)$ is the source signal. We assume that the received source

[1] Beamformers are spatial filters, therefore the terms *beamformer* and *filter* will be used interchangeably in this chapter.

[2] The dependency on time is omitted for brevity. In practice, the signals acquired using a spherical microphone array are usually processed in the short-time Fourier transform domain, as explained in Sect. 3.1, where the discrete frequency index is denoted by ν.

D. Jarrett et al., *Theory and Applications of Spherical Microphone Array Processing*, Springer Topics in Signal Processing 9,
DOI 10.1007/978-3-319-42211-4_7

signals X and noise signals V are mutually uncorrelated. The received source signals X originate from a single source, and are therefore, by definition, coherent across the array.

When using spherical microphone arrays, it is convenient to work in the spherical harmonic domain (SHD) [28, 31], instead of the spatial domain. In this chapter, we assume error-free spatial sampling, and refer the reader to Chap. 3 for information on spatial sampling and aliasing. By applying the complex spherical harmonic transform (SHT)[3] to the signal model in (7.1), we obtain the SHD signal model

$$P_{lm}(k) = H_{lm}(k)S(k) + V_{lm}(k) \tag{7.2a}$$

$$= X_{lm}(k) + V_{lm}(k), \tag{7.2b}$$

where $P_{lm}(k)$, $X_{lm}(k)$, $H_{lm}(k)$ and $V_{lm}(k)$ are respectively the spherical harmonic transforms of the signals $P(k, \mathbf{r}_q)$, $X(k, \mathbf{r}_q)$, $H(k, \mathbf{r}_q)$ and $V(k, \mathbf{r}_q)$, as defined in (3.6), and are referred to as *eigenbeams* to reflect the fact that the spherical harmonics are eigensolutions of the wave equation in spherical coordinates [34]. The order and degree of the spherical harmonics are respectively denoted as l and m.

The eigenbeams $P_{lm}(k)$, $X_{lm}(k)$, $H_{lm}(k)$ and $V_{lm}(k)$ are functions of the frequency-dependent mode strength $b_l(k)$, which is in turn a function of the array properties such as radius, microphone type and configuration. Mode strength expressions for two common types of arrays, the open and rigid arrays with omnidirectional microphones, are given in Sect. 3.4.2. To remove the dependence of the eigenbeams on the mode strength, we divide the eigenbeams by the mode strength (as in [30]), thus giving mode strength compensated eigenbeams, and the signal model is then written as

$$\widetilde{P}_{lm}(k) = \left[\sqrt{4\pi} b_l(k)\right]^{-1} P_{lm}(k) \tag{7.3a}$$

$$= \widetilde{H}_{lm}(k)S(k) + \widetilde{V}_{lm}(k) \tag{7.3b}$$

$$= \widetilde{X}_{lm}(k) + \widetilde{V}_{lm}(k), \tag{7.3c}$$

where $\widetilde{P}_{lm}(k)$, $\widetilde{H}_{lm}(k)$, $\widetilde{X}_{lm}(k)$ and $\widetilde{V}_{lm}(k)$ respectively denote the eigenbeams $P_{lm}(k)$, $H_{lm}(k)$, $X_{lm}(k)$ and $V_{lm}(k)$ after mode strength compensation.

The design of a beamformer involves the choice of a desired signal, which the beamformer will seek to estimate. The desired signal is commonly chosen as the source component $X(k, \mathbf{r}_{\text{ref}})$ of the signal $P(k, \mathbf{r}_{\text{ref}})$ received at a particular reference microphone with position $\mathbf{r}_{\text{ref}}$ [2, 13, 17]. When working with linear arrays of omnidirectional microphones, the choice of this reference microphone is not usually very important, since the power of the desired signal is likely to be similar at all microphones in a particular array. With spherical arrays, however, this is no longer necessarily the case. In particular, with a rigid array, due to the scattering effects of

[3]If the real SHT is applied instead of the complex SHT, the complex spherical harmonics Y_{lm} used throughout this chapter should be replaced with the real spherical harmonics R_{lm}, as defined in Sect. 3.3.

the sphere, the microphones have some added directionality, and the power on the occluded side of the array could be lower.

For this reason, we choose as a reference a virtual *omnidirectional* microphone $\mathcal{M}_{\text{ref}}$, placed at the centre of the sphere, which is also the origin of the spherical coordinate system employed. The signal $\widetilde{P}_{00}(k)$ is equal to the signal that would be received by the reference microphone $\mathcal{M}_{\text{ref}}$, if the sphere were not present, as shown in the Appendix of Chap. 5. Our aim is then to estimate the received source component $\widetilde{X}_{00}(k)$ of this signal using a beamformer.

For convenience, we rewrite the signal model (7.3) in vector notation, where the vectors all have length $N = (L + 1)^2$, the total number of eigenbeams from order $l = 0$ to $l = L$:

$$\begin{aligned}\widetilde{\mathbf{p}}(k) &= \widetilde{\mathbf{h}}(k)S(k) + \widetilde{\mathbf{v}}(k) &\text{(7.4a)}\\ &= \widetilde{\mathbf{x}}(k) + \widetilde{\mathbf{v}}(k) &\text{(7.4b)}\\ &= \mathbf{d}(k)\widetilde{X}_{00}(k) + \widetilde{\mathbf{v}}(k), &\text{(7.4c)}\end{aligned}$$

where

$$\widetilde{\mathbf{p}}(k) = \left[\widetilde{P}_{00}(k)\, \widetilde{P}_{1(-1)}(k)\, \widetilde{P}_{10}(k)\, \widetilde{P}_{11}(k)\, \widetilde{P}_{2(-2)}(k) \cdots \widetilde{P}_{LL}(k)\right]^{\mathrm{T}}, \tag{7.5}$$

$$\widetilde{\mathbf{h}}(k) = \left[\widetilde{H}_{00}(k)\, \widetilde{H}_{1(-1)}(k)\, \widetilde{H}_{10}(k)\, \widetilde{H}_{11}(k)\, \widetilde{H}_{2(-2)}(k) \cdots \widetilde{H}_{LL}(k)\right]^{\mathrm{T}}, \tag{7.6}$$

$$\begin{aligned}\mathbf{d}(k) &= \left[1\; \frac{\widetilde{H}_{1(-1)}(k)}{\widetilde{H}_{00}(k)}\; \frac{\widetilde{H}_{10}(k)}{\widetilde{H}_{00}(k)}\; \frac{\widetilde{H}_{11}(k)}{\widetilde{H}_{00}(k)}\; \frac{\widetilde{H}_{2(-2)}(k)}{\widetilde{H}_{00}(k)} \cdots \frac{\widetilde{H}_{LL}(k)}{\widetilde{H}_{00}(k)}\right]^{\mathrm{T}}\\ &= \left[D_{00}(k)\, D_{1(-1)}(k)\, D_{10}(k)\, D_{11}(k)\, D_{2(-2)}(k) \cdots D_{LL}(k)\right]^{\mathrm{T}}, \end{aligned}\tag{7.7}$$

$(\cdot)^{\mathrm{T}}$ denotes the vector transpose, and $\widetilde{\mathbf{x}}(k)$ and $\widetilde{\mathbf{v}}(k)$ are defined similarly to $\widetilde{\mathbf{p}}(k)$. The vector $\mathbf{d}$ is commonly referred to as the relative transfer function (RTF) or propagation vector [12]. In the following, it is assumed that $H_{00}(k) \neq 0\, \forall k$, such that the RTF vector $\mathbf{d}(k)$ is always defined.

As $X(k, \mathbf{r}_q)$ and $V(k, \mathbf{r}_q)$ are mutually uncorrelated, and the SHT and division by the mode strength are linear operations, $\widetilde{X}_{lm}(k)$ and $\widetilde{V}_{lm}(k)$ are also mutually uncorrelated. The power spectral density (PSD) matrix $\boldsymbol{\Phi}_{\widetilde{\mathbf{p}}}$ of $\widetilde{\mathbf{p}}$ can therefore be expressed as

$$\begin{aligned}\boldsymbol{\Phi}_{\widetilde{\mathbf{p}}}(k) &= \mathrm{E}\left\{\widetilde{\mathbf{p}}(k)\widetilde{\mathbf{p}}^{\mathrm{H}}(k)\right\}\\ &= \boldsymbol{\Phi}_{\widetilde{\mathbf{x}}}(k) + \boldsymbol{\Phi}_{\widetilde{\mathbf{v}}}(k), \end{aligned}\tag{7.8}$$

where

$$\begin{aligned}\boldsymbol{\Phi}_{\widetilde{\mathbf{x}}}(k) &= \mathrm{E}\left\{\widetilde{\mathbf{x}}(k)\widetilde{\mathbf{x}}^{\mathrm{H}}(k)\right\} = \phi_{\widetilde{X}_{00}}(k)\mathbf{d}(k)\mathbf{d}^{\mathrm{H}}(k) \text{ and}\\ \boldsymbol{\Phi}_{\widetilde{\mathbf{v}}}(k) &= \mathrm{E}\left\{\widetilde{\mathbf{v}}(k)\widetilde{\mathbf{v}}^{\mathrm{H}}(k)\right\}\end{aligned}$$

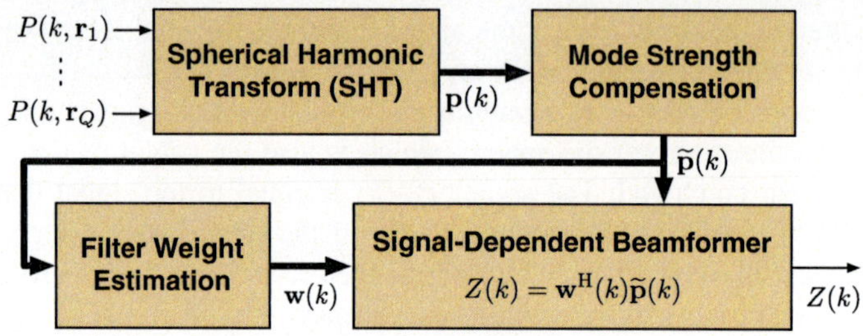

Fig. 7.1 Block diagram of a signal-dependent beamformer

are respectively the PSD matrices of $\widetilde{\mathbf{x}}(k)$ and $\widetilde{\mathbf{v}}(k)$, $\phi_{\widetilde{X}_{00}}(k) = \mathrm{E}\left\{\left|\widetilde{X}_{00}(k)\right|^2\right\}$ is the variance of $\widetilde{X}_{00}(k)$, and $(\cdot)^{\mathrm{H}}$ denotes the Hermitian transpose.

As in Chap. 6, the output $Z(k)$ of our beamformer is obtained by applying a complex weight to each eigenbeam, and summing over all eigenbeams (a *filter-and-sum* operation)[4]:

$$\begin{aligned} Z(k) &= \mathbf{w}^{\mathrm{H}}(k)\widetilde{\mathbf{p}}(k) \\ &= \mathbf{w}^{\mathrm{H}}(k)\widetilde{\mathbf{x}}(k) + \mathbf{w}^{\mathrm{H}}(k)\widetilde{\mathbf{v}}(k) \\ &= \widetilde{X}_{\mathrm{f}}(k) + \widetilde{V}_{\mathrm{r}}(k), \end{aligned} \tag{7.9}$$

where $\widetilde{X}_{\mathrm{f}}(k) = \mathbf{w}^{\mathrm{H}}(k)\widetilde{\mathbf{x}}(k) = \mathbf{w}^{\mathrm{H}}(k)\mathbf{d}(k)\widetilde{X}_{00}(k)$ is the filtered desired signal and $\widetilde{V}_{\mathrm{r}}(k) = \mathbf{w}^{\mathrm{H}}(k)\widetilde{\mathbf{v}}(k)$ is the residual noise. The filtered desired and residual noise signals are mutually uncorrelated; therefore, the variance of $Z(k)$ is given by the sum of two variances

$$\begin{aligned} \phi_Z(k) &= \mathbf{w}^{\mathrm{H}}(k)\mathbf{\Phi}_{\widetilde{\mathbf{p}}}(k)\mathbf{w}(k) \\ &= \phi_{\widetilde{X}_{\mathrm{f}}}(k) + \phi_{\widetilde{V}_{\mathrm{r}}}(k), \end{aligned} \tag{7.10}$$

where $\phi_{\widetilde{X}_{\mathrm{f}}}(k) = \phi_{\widetilde{X}_{00}}(k)\left|\mathbf{w}^{\mathrm{H}}(k)\mathbf{d}(k)\right|^2$ and $\phi_{\widetilde{V}_{\mathrm{r}}}(k) = \mathbf{w}^{\mathrm{H}}(k)\mathbf{\Phi}_{\widetilde{\mathbf{v}}}(k)\mathbf{w}(k)$.

The beamforming process is illustrated in Fig. 7.1. It begins by capturing the spatial domain sound pressure signals $P(k, \mathbf{r}_q)$ at microphones $q \in \{1, \dots, Q\}$, and applying the SHT to obtain a vector of SHD sound pressure signals or *eigenbeams* $\mathbf{p}(k)$. We then divide the eigenbeams by the mode strength, yielding a vector $\widetilde{\mathbf{p}}(k)$. Finally, the eigenbeams $\widetilde{\mathbf{p}}(k)$ are weighted and summed to obtain the beamformer output $Z(k)$. The weights $\mathbf{w}(k)$ of the signal-dependent beamformer, which are a function of the eigenbeams $\widetilde{\mathbf{p}}(k)$, are chosen in order to achieve certain performance objectives, which will be defined in the following section.

---

[4]We use the complex conjugate weights $\mathbf{w}^{\mathrm{H}}$ rather than the weights $\mathbf{w}^{\mathrm{T}}$; this notational convention originates in the spatial domain [37].

7.2 Performance Measures

In this section, we define a number of signal-dependent performance measures that can be used to design and evaluate the beamformers to be presented in Sect. 7.3.

The performance measures that are defined in this section are computed for a particular wavenumber k, and are therefore referred to as *subband* measures. In contrast, *full-band* measures would be computed over all wavenumbers.

7.2.1 *Speech Distortion Index*

The filtering process that is intrinsic to beamforming may, in some cases, unfortunately introduce distortion into the desired signal. The *speech distortion index* υ_{sd} measures this distortion by computing the normalized mean square error in the frequency domain between the filtered desired signal $\widetilde{X}_{\mathrm{f}}(k)$ and the desired signal $\widetilde{X}_{00}(k)$ [4, 19], such that

$$\upsilon_{\mathrm{sd}}\left[\mathbf{w}(k)\right] = \frac{\mathrm{E}\left\{\left|\widetilde{X}_{\mathrm{f}}(k) - \widetilde{X}_{00}(k)\right|^2\right\}}{\phi_{\widetilde{X}_{00}}(k)} \tag{7.11a}$$

$$= \left|\mathbf{w}^{\mathrm{H}}(k)\mathbf{d}(k) - 1\right|^2. \tag{7.11b}$$

It is clear that the weights of a beamformer that does not distort the desired signal, or in other words, weights that give rise to a speech distortion index of 0, must satisfy the *distortionless constraint*

$$\mathbf{w}^{\mathrm{H}}(k)\mathbf{d}(k) = 1, \forall k, \tag{7.12}$$

which we will make use of in the design of the minimum variance distortionless response (MVDR) beamformer in Sect. 7.3.3.

The speech distortion index should be as low as possible for good quality speech and is normally upper bounded by 1. A speech distortion index of around −20–−10 dB is considered to be low, while a high speech distortion index, typically in the range −5–0 dB, will normally result in obvious distortions. While in the single-channel case any noise reduction comes at the expense of speech distortion [33], in the multichannel case the distortion can in theory be eliminated entirely [3].

For the purposes of a performance evaluation, the speech distortion index is typically computed using short, 10–20 ms time frames [22], and then averaged over all frames that contain speech; it is then referred to as the *segmental* speech distortion index.

7.2.2 Noise Reduction Factor

The *noise reduction factor* quantifies the reduction in noise power due to the beamformer, defined as the ratio of the power of the noise at the beamformer input $\widetilde{V}_{00}(k)$ to the power of the residual noise at the beamformer output $\widetilde{V}_{\mathrm{r}}(k)$ [2, 15]:

$$\xi_{\mathrm{nr}}[\mathbf{w}(k)] = \frac{\phi_{\widetilde{V}_{00}}(k)}{\phi_{\widetilde{V}_{\mathrm{r}}}(k)} \tag{7.13a}$$

$$= \frac{\phi_{\widetilde{V}_{00}}(k)}{\mathbf{w}^{\mathrm{H}}(k)\boldsymbol{\Phi}_{\widetilde{\mathbf{v}}}(k)\mathbf{w}(k)}, \tag{7.13b}$$

where $\phi_{\widetilde{V}_{00}}(k) = \mathrm{E}\left\{\left|\widetilde{V}_{00}(k)\right|^2\right\}$ is the variance of $\widetilde{V}_{00}(k)$.

This noise reduction factor is defined with respect to the reference microphone $\mathcal{M}_{\mathrm{ref}}$. It should however be noted that:

- The reference microphone $\mathcal{M}_{\mathrm{ref}}$ is omnidirectional. On the other hand, as noted in Sect. 7.1, with a rigid array, the sphere provides the array's microphones with some directionality, even if the array is constructed of omnidirectional microphones. Consequently, the power of *spatially coherent* noise may well be lower at one of these microphones than at $\mathcal{M}_{\mathrm{ref}}$. However, the microphone with the lowest coherent noise level will vary, and it is therefore more convenient to use a single (omnidirectional) reference microphone.
- Assuming the array's Q microphones are uniformly distributed on the sphere, the power of *spatially incoherent* noise (such as sensor noise) at $\mathcal{M}_{\mathrm{ref}}$ is reduced by a factor of $Q\,|b_0(k)|^2$ with respect to its power at the microphones [20]. The Q factor is present because the reference microphone signal $\widetilde{P}_{00}(k)$ is formed using all Q individual microphone signals, and spatially incoherent noise sums destructively. The mode strength compensation operation accounts for the $|b_0(k)|^2$ factor. For an open or rigid array of omnidirectional microphones, at low frequencies where $b_0(k) \approx 1$, the power of incoherent noise is Q times smaller at $\mathcal{M}_{\mathrm{ref}}$ than at the microphones (for example, 15 dB lower for $Q = 32$ microphones).

The noise reduction factor only provides useful performance information when viewed alongside a performance measure that relates to the desired speech, such as the speech distortion index, since a trivial filter $\mathbf{w}(k) = \mathbf{0}_N$ (where $\mathbf{0}_N$ is a column vector of N zeros) would achieve an infinite noise reduction factor, yet it would not be of any practical use.

When evaluating the performance of a beamformer, this measure is typically computed using short, 10–20 ms time frames, and then averaged in the logarithm domain over all time frames; it is then referred to as the *segmental* noise reduction factor.

7.2.3 Array Gain

The *array gain* is the signal-to-noise ratio (SNR) improvement obtained using the beamformer [35], that is, the ratio of the output SNR, oSNR, to the input SNR, iSNR, given by

$$\mathcal{A}\left[\mathbf{w}(k)\right] = \frac{\text{oSNR}\left[\mathbf{w}(k)\right]}{\text{iSNR}(k)} = \frac{\phi_{\widetilde{X}_{\text{f}}}(k)}{\phi_{\widetilde{V}_{\text{r}}}(k)} \frac{\phi_{\widetilde{V}_{00}}(k)}{\phi_{\widetilde{X}_{00}}(k)} \tag{7.14a}$$

$$= \frac{\phi_{\widetilde{V}_{00}}(k) \left|\mathbf{w}^{\text{H}}(k)\mathbf{d}(k)\right|^2}{\mathbf{w}^{\text{H}}(k)\mathbf{\Phi}_{\widetilde{\mathbf{v}}}(k)\mathbf{w}(k)}. \tag{7.14b}$$

The array gain is also defined with respect to the reference microphone $\mathcal{M}_{\text{ref}}$; the consequences of this definition are discussed in Sect. 7.2.2.

The performance of a beamformer is frequently evaluated with the *segmental* array gain, which involves computing the array gain using short, 10–20 ms time frames, and then averaging it in the logarithm domain over all frames that contain speech. These *active* frames can be determined using ITU-T Rec. P.56 [18], for example.

7.2.4 Mean Square Error

Due to its simplicity, one of the most frequently used criteria for designing optimal beamformers is the *mean square error* (MSE) criterion. The error between the beamformer output signal $Z(k)$ and the desired signal $\widetilde{X}_{00}(k)$ is given by

$$\begin{aligned}
\text{E}(k) &= Z(k) - \widetilde{X}_{00}(k) \\
&= \mathbf{w}^{\text{H}}(k)\widetilde{\mathbf{p}}(k) - \widetilde{X}_{00}(k) \\
&= \mathbf{w}^{\text{H}}(k)\widetilde{\mathbf{x}}(k) - \widetilde{X}_{00}(k) + \mathbf{w}^{\text{H}}(k)\widetilde{\mathbf{v}}(k) \\
&= \left[\mathbf{w}^{\text{H}}(k)\mathbf{d}(k) - 1\right]\widetilde{X}_{00}(k) + \mathbf{w}^{\text{H}}(k)\widetilde{\mathbf{v}}(k).
\end{aligned} \tag{7.15}$$

The MSE is then

$$\begin{aligned}
J\left[\mathbf{w}(k)\right] &= \text{E}\left\{|\text{E}(k)|^2\right\} \\
&= \phi_{\widetilde{X}_{00}}(k) \left|\mathbf{w}^{\text{H}}(k)\mathbf{d}(k) - 1\right|^2 + \phi_{\widetilde{V}_{\text{r}}}(k).
\end{aligned} \tag{7.16}$$

Using (7.11b) and (7.13a), the MSE can be expressed as a function of two other performance measures:

$$J\left[\mathbf{w}(k)\right] = \phi_{\widetilde{X}_{00}}(k) \upsilon_{\text{sd}}\left[\mathbf{w}(k)\right] + \frac{\phi_{\widetilde{V}_{00}}(k)}{\xi_{\text{nr}}\left[\mathbf{w}(k)\right]}. \tag{7.17}$$

The MSE is an increasing function of the speech distortion index υ_{sd}, and a decreasing function of the noise reduction factor ξ_{nr}. Minimizing the MSE, as we will do in Sect. 7.3.2, is equivalent to minimizing jointly the speech distortion index and maximizing the noise reduction factor.

7.3 Signal-Dependent Beamformers

In this section, we present a number of signal-dependent beamformers, designed based on the performance measures in Sect. 7.2. They are similar to beamformers commonly used in the spatial domain but are formulated in the SHD with a reference microphone $\mathcal{M}_{\text{ref}}$.

All the filters presented in this section, with the exception of the linearly constrained minimum variance (LCMV) filter, maximize the subband output SNR. However, depending on the design criteria, the full-band output SNR may be different.

7.3.1 Maximum SNR Filter

The SNR at the output of a beamformer with weights $\mathbf{w}(k)$ is given by

$$\text{oSNR}\left[\mathbf{w}(k)\right] = \frac{\phi_{\widetilde{X}_{\text{f}}}(k)}{\phi_{\widetilde{V}_{\text{r}}}(k)} \tag{7.18a}$$

$$= \frac{\phi_{\widetilde{X}_{00}}(k)\left|\mathbf{w}^{\text{H}}(k)\mathbf{d}(k)\right|^2}{\mathbf{w}^{\text{H}}(k)\mathbf{\Phi}_{\widetilde{\mathbf{v}}}(k)\mathbf{w}(k)}. \tag{7.18b}$$

The filter that maximizes the subband output SNR, referred to as a *maximum* SNR filter, can then be determined as [2]

$$\begin{aligned}\mathbf{w}_{\text{max}}(k) &= \underset{\mathbf{w}(k)}{\arg\max}\ \text{oSNR}\left[\mathbf{w}(k)\right] \\ &= \alpha(k)\mathbf{\Phi}_{\widetilde{\mathbf{v}}}^{-1}(k)\mathbf{d}(k),\end{aligned} \tag{7.19}$$

where $\alpha(k) \neq 0$ is an arbitrary frequency-dependent scaling factor. The Wiener, MVDR and parametric Wiener filters presented in the sections that follow are all equal to the maximum SNR filter for a specific choice of this frequency-dependent scaling factor.

In (7.18b), we recognize the generalized Rayleigh quotient. Since this quotient is maximized by the maximum eigenvector (i.e., the eigenvector associated with the largest eigenvalue) of the matrix $\mathbf{\Phi}_{\widetilde{\mathbf{v}}}^{-1}(k)\mathbf{\Phi}_{\widetilde{\mathbf{x}}}(k)$, this maximum eigenvector is also a maximum SNR filter [2]. Alternatively, the filter is given by the generalized eigenvectors associated with the largest generalized eigenvalues of the matrix pencils $(\mathbf{\Phi}_{\widetilde{\mathbf{x}}}(k), \mathbf{\Phi}_{\widetilde{\mathbf{v}}}(k))$ and $(\mathbf{\Phi}_{\widetilde{\mathbf{p}}}(k), \mathbf{\Phi}_{\widetilde{\mathbf{v}}}(k))$.

7.3.2 *Wiener Filter*

The *Wiener filter* [38] minimizes the MSE defined in (7.16):

$$\mathbf{w}_{\mathrm{W}}(k) = \underset{\mathbf{w}(k)}{\arg\min}\, J\left[\mathbf{w}(k)\right]. \tag{7.20}$$

It can be derived by expressing the MSE as

$$\begin{aligned} J\left[\mathbf{w}(k)\right] &= \mathrm{E}\left\{|\mathrm{E}(k)|^2\right\} \\ &= \mathrm{E}\left\{\left(\mathbf{w}^{\mathrm{H}}(k)\widetilde{\mathbf{p}}(k) - \widetilde{X}_{00}(k)\right)\left(\mathbf{w}^{\mathrm{H}}(k)\widetilde{\mathbf{p}}(k) - \widetilde{X}_{00}(k)\right)^{\mathrm{H}}\right\} \\ &= \mathbf{w}^{\mathrm{H}}(k)\mathbf{\Phi}_{\widetilde{\mathbf{p}}}(k)\mathbf{w}(k) + \phi_{\widetilde{X}_{00}}(k) - \mathbf{w}^{\mathrm{H}}(k)\mathrm{E}\left\{\widetilde{\mathbf{p}}(k)\widetilde{X}_{00}^{*}(k)\right\} \\ &\qquad - \mathrm{E}\left\{\widetilde{\mathbf{p}}^{\mathrm{H}}(k)\widetilde{X}_{00}(k)\right\}\mathbf{w}(k). \end{aligned} \tag{7.21}$$

The Wiener filter weights $\mathbf{w}_{\mathrm{W}}(k)$ must then satisfy

$$\begin{aligned} \nabla_{\mathbf{w}_{\mathrm{W}}^{*}} J\left[\mathbf{w}_{\mathrm{W}}(k)\right] &= \mathbf{0}_N \\ \mathbf{\Phi}_{\widetilde{\mathbf{p}}}(k)\mathbf{w}_{\mathrm{W}}(k) - \mathrm{E}\left\{\widetilde{\mathbf{p}}(k)\widetilde{X}_{00}^{*}(k)\right\} &= \mathbf{0}_N \\ \mathbf{\Phi}_{\widetilde{\mathbf{p}}}(k)\mathbf{w}_{\mathrm{W}}(k) - \mathrm{E}\left\{\widetilde{\mathbf{x}}(k)\widetilde{X}_{00}^{*}(k)\right\} - \mathrm{E}\left\{\widetilde{\mathbf{v}}(k)\widetilde{X}_{00}^{*}(k)\right\} &= \mathbf{0}_N, \end{aligned} \tag{7.22}$$

where $\mathbf{0}_N$ is a column vector of N zeros and $\nabla_{\mathbf{w}_{\mathrm{W}}^{*}} J\left[\mathbf{w}_{\mathrm{W}}(k)\right]$ is the complex gradient vector of $J\left[\mathbf{w}_{\mathrm{W}}(k)\right]$ with respect to $\mathbf{w}_{\mathrm{W}}^{*}$, as defined in [7]. As $\widetilde{X}_{lm}(k)$ and $\widetilde{V}_{lm}(k)$ are mutually uncorrelated, $\mathrm{E}\left\{\widetilde{\mathbf{v}}(k)\widetilde{X}_{00}^{*}(k)\right\} = \mathbf{0}_N$, and hence we find

$$\begin{aligned} \mathbf{w}_{\mathrm{W}}(k) &= \mathbf{\Phi}_{\widetilde{\mathbf{p}}}^{-1}(k)\mathrm{E}\left\{\widetilde{\mathbf{x}}(k)\widetilde{X}_{00}^{*}(k)\right\} \\ &= \phi_{\widetilde{X}_{00}}(k)\mathbf{\Phi}_{\widetilde{\mathbf{p}}}^{-1}(k)\mathbf{d}(k). \end{aligned} \tag{7.23}$$

Since $\mathbf{\Phi}_{\widetilde{\mathbf{x}}}(k) = \phi_{\widetilde{X}_{00}}(k)\mathbf{d}(k)\mathbf{d}^{\mathrm{H}}(k)$, the filter weights can also be expressed as [2]

$$\mathbf{w}_{\mathrm{W}}(k) = \mathbf{\Phi}_{\widetilde{\mathbf{p}}}^{-1}(k)\mathbf{\Phi}_{\widetilde{\mathbf{x}}}(k)\mathbf{i}_N \tag{7.24a}$$

$$= \left[\mathbf{I}_{N\times N} - \mathbf{\Phi}_{\widetilde{\mathbf{p}}}^{-1}(k)\mathbf{\Phi}_{\widetilde{\mathbf{v}}}(k)\right]\mathbf{i}_N, \tag{7.24b}$$

where $\mathbf{I}_{N\times N}$ denotes an $N \times N$ identity matrix, and $\mathbf{i}_N$ denotes its first column. In (7.24b), the weights are only a function of the second-order statistics of the noise and observation signals.

It can also be shown [2] that (7.23) can be expressed as

$$\mathbf{w}_{\mathrm{W}}(k) = \frac{\phi_{\tilde{X}_{00}}(k)\boldsymbol{\Phi}_{\tilde{\mathbf{v}}}^{-1}(k)\mathbf{d}(k)}{1 + \phi_{\tilde{X}_{00}}(k)\mathbf{d}^{\mathrm{H}}(k)\boldsymbol{\Phi}_{\tilde{\mathbf{v}}}^{-1}(k)\mathbf{d}(k)}. \tag{7.25}$$

We then see that the Wiener filter is a special case of the maximum SNR filter in (7.19) for the case where the weights of the maximum SNR filter are computed using the scaling factor

$$\alpha(k) = \frac{\phi_{\tilde{X}_{00}}(k)}{1 + \phi_{\tilde{X}_{00}}(k)\mathbf{d}^{\mathrm{H}}(k)\boldsymbol{\Phi}_{\tilde{\mathbf{v}}}^{-1}(k)\mathbf{d}(k)}. \tag{7.26}$$

7.3.3 Minimum Variance Distortionless Response Filter

The MVDR beamformer or *Capon beamformer* [10] minimizes the residual noise power (or equivalently, maximizes the noise reduction factor) while imposing a distortionless constraint on the desired signal:

$$\begin{aligned} \min_{\mathbf{w}(k)} \phi_{\tilde{V}_{\mathrm{r}}}(k) \quad &\text{subject to} \quad \upsilon_{\mathrm{sd}}\left[\mathbf{w}(k)\right] = 0 \\ \min_{\mathbf{w}(k)} \mathbf{w}^{\mathrm{H}}(k)\boldsymbol{\Phi}_{\tilde{\mathbf{v}}}(k)\mathbf{w}(k) \quad &\text{subject to} \quad \mathbf{w}^{\mathrm{H}}(k)\mathbf{d}(k) = 1. \end{aligned} \tag{7.27}$$

Following the approach proposed by Brandwood [7], if we use a Lagrange multiplier to adjoin the constraint to the cost function, the MVDR filter is then given by

$$\mathbf{w}_{\mathrm{MVDR}}(k) = \arg\min_{\mathbf{w}(k)} \mathcal{L}(\mathbf{w}(k), \lambda), \tag{7.28}$$

where $\mathcal{L}$ is the complex Lagrangian given by

$$\begin{aligned} \mathcal{L}(\mathbf{w}(k), \lambda) = \mathbf{w}^{\mathrm{H}}(k)\boldsymbol{\Phi}_{\tilde{\mathbf{v}}}(k)\mathbf{w}(k) &+ \lambda\left(\mathbf{w}^{\mathrm{H}}(k)\mathbf{d}(k) - 1\right) \\ &+ \lambda^*\left(\mathbf{d}^{\mathrm{H}}(k)\mathbf{w}(k) - 1\right) \end{aligned} \tag{7.29}$$

and λ is the Lagrange multiplier. Setting the gradient of $\mathcal{L}(\mathbf{w}_{\mathrm{MVDR}}(k), \lambda)$ with respect to $\mathbf{w}_{\mathrm{MVDR}}^*$ to 0 yields

$$\begin{aligned} \nabla_{\mathbf{w}_{\mathrm{MVDR}}^*}\mathcal{L}(\mathbf{w}_{\mathrm{MVDR}}(k), \lambda) &= \mathbf{0}_N \\ \boldsymbol{\Phi}_{\tilde{\mathbf{v}}}(k)\mathbf{w}_{\mathrm{MVDR}}(k) + \lambda\mathbf{d}(k) &= \mathbf{0}_N. \end{aligned} \tag{7.30}$$

Using the constraint in (7.27), we then find [2, 21]

$$\mathbf{w}_{\mathrm{MVDR}}(k) = \frac{\mathbf{\Phi}_{\tilde{\mathbf{v}}}^{-1}(k)\mathbf{d}(k)}{\mathbf{d}^{\mathrm{H}}(k)\mathbf{\Phi}_{\tilde{\mathbf{v}}}^{-1}(k)\mathbf{d}(k)}. \tag{7.31}$$

The MVDR filter is a special case of the maximum SNR filter in (7.19) for the case where the weights of the maximum SNR filter are computed using the scaling factor

$$\alpha(k) = \frac{1}{\mathbf{d}^{\mathrm{H}}(k)\mathbf{\Phi}_{\tilde{\mathbf{v}}}^{-1}(k)\mathbf{d}(k)}. \tag{7.32}$$

Instead of minimizing the noise power $\phi_{\tilde{V}_{\mathrm{r}}}$ at the output of the beamformer, one can minimize the total power of the beamformer's output (i.e., desired speech plus residual noise). This optimization problem can be written as

$$\min_{\mathbf{w}(k)} \mathbf{w}^{\mathrm{H}}(k)\mathbf{\Phi}_{\tilde{\mathbf{p}}}(k)\mathbf{w}(k) \quad \text{subject to} \quad \mathbf{w}^{\mathrm{H}}(k)\mathbf{d}(k) = 1, \tag{7.33}$$

and its solution is the minimum power distortionless response (MPDR) filter

$$\mathbf{w}_{\mathrm{MPDR}}(k) = \frac{\mathbf{\Phi}_{\tilde{\mathbf{p}}}^{-1}(k)\mathbf{d}(k)}{\mathbf{d}^{\mathrm{H}}(k)\mathbf{\Phi}_{\tilde{\mathbf{p}}}^{-1}(k)\mathbf{d}(k)}. \tag{7.34}$$

As we do not have access to the true RTF vector $\mathbf{d}(k)$, we need an estimate of $\mathbf{d}(k)$ to compute the MVDR and MPDR filters. It can be shown that the MVDR and MPDR filters are equivalent only when the estimated and true RTF vectors are equal. The MPDR filter is, however, known to be sensitive to RTF estimation errors. When the solution is obtained using an adaptive solution, severe signal cancellation can occur.

Relationship to Signal-Independent Beamformers

Under specific assumptions about the spatial characteristics of the desired speech and noise signals, the MVDR beamformer can be equivalent to beamformers introduced in Chap. 6. In this section, we derive two such equivalences. The two types of noise field we deal with are the *spherically isotropic noise field*, where the noise is spatially diffuse [24], and the *spatially white noise field*, where the noise is spatially incoherent, that is, the noise signals $V(k, \mathbf{r}_q)$ at each microphone are mutually uncorrelated.

Property 7.1 *In a spherically isotropic noise field, assuming an anechoic environment and plane-wave incidence, the MVDR beamformer reduces to a signal-independent maximum directivity beamformer, as introduced in Sect. 6.3.1.1.*

Proof In a spherically isotropic noise field, the noise PSD matrix (after mode strength compensation) is given by [8, 23, 40]

$$\mathbf{\Phi}_{\tilde{\mathbf{v}}}(k) = \sigma_v^2(k)\mathbf{I}_{N\times N}. \tag{7.35}$$

By substituting (7.35) into (7.31), we find

$$\begin{aligned}\mathbf{w}_{\mathrm{MVDR}}(k) &= \frac{\sigma_v^{-2}(k)\mathbf{I}_{N\times N}\mathbf{d}(k)}{\mathbf{d}^{\mathrm{H}}(k)\sigma_v^{-2}(k)\mathbf{I}_{N\times N}\mathbf{d}(k)}\\ &= \frac{\mathbf{d}(k)}{\mathbf{d}^{\mathrm{H}}(k)\mathbf{d}(k)}.\end{aligned} \tag{7.36}$$

In an anechoic environment, assuming plane-wave incidence from a direction Ω_{s}, the RTF vector $\mathbf{d}(k)$ is given by [8]

$$\mathbf{d}(k) = \frac{\mathbf{y}^*(\Omega_{\mathrm{s}})}{Y_{00}^*(\Omega_{\mathrm{s}})}, \tag{7.37}$$

where $Y_{lm}(\Omega_{\mathrm{s}})$ denotes the complex spherical harmonic of order l and degree m evaluated at an angle Ω_{s}, as defined in (2.14), and

$$\mathbf{y}(\Omega_{\mathrm{s}}) = \left[Y_{00}(\Omega_{\mathrm{s}})\, Y_{1(-1)}(\Omega_{\mathrm{s}})\, Y_{10}(\Omega_{\mathrm{s}})\, Y_{11}(\Omega_{\mathrm{s}}) \cdots Y_{LL}(\Omega_{\mathrm{s}})\right]^{\mathrm{T}}, \tag{7.38}$$

and therefore, using $Y_{00}(\cdot) = 1/\sqrt{4\pi}$ and the spherical harmonic addition theorem [39], the beamformer weights simplify to

$$\begin{aligned}\mathbf{w}_{\mathrm{MVDR}}(k) &= \frac{\mathbf{y}^*(\Omega_{\mathrm{s}})}{\sqrt{4\pi}\,\mathbf{y}^{\mathrm{T}}(\Omega_{\mathrm{s}})\mathbf{y}^*(\Omega_{\mathrm{s}})}\\ &= \frac{\mathbf{y}^*(\Omega_{\mathrm{s}})}{\sqrt{4\pi}\sum_{l=0}^{L}\sum_{m=-l}^{l}\left|Y_{lm}^*(\Omega_{\mathrm{s}})\right|^2}\\ &= \frac{4\pi\mathbf{y}^*(\Omega_{\mathrm{s}})}{\sqrt{4\pi}\sum_{l=0}^{L}(2l+1)}\\ &= \frac{\sqrt{4\pi}}{(L+1)^2}\mathbf{y}^*(\Omega_{\mathrm{s}}).\end{aligned} \tag{7.39}$$

Finally the beamformer output is given by

$$\begin{aligned}Z(k) &= \frac{\sqrt{4\pi}}{N}\sum_{l=0}^{L}\sum_{m=-l}^{l} Y_{lm}(\Omega_{\mathrm{s}})\widetilde{P}_{lm}(k)\\ &= \frac{1}{N}\sum_{l=0}^{L}\sum_{m=-l}^{l}\frac{Y_{lm}(\Omega_{\mathrm{s}})}{b_l(k)}P_{lm}(k),\end{aligned} \tag{7.40}$$

which is proportional to the output of a maximum directivity beamformer, as defined in (6.36), and therefore Property 7.1 holds.

Property 7.2 *In a spatially white noise field, assuming an anechoic environment and plane-wave incidence, the MVDR beamformer reduces to a signal-independent maximum white noise gain (WNG) beamformer, as introduced in Sect. 6.3.1.2.*

Proof In a spatially white noise field, the noise PSD matrix (after mode strength compensation) is given by [20]

$$\mathbf{\Phi}_{\tilde{\mathbf{v}}}(k) = \sigma_v^2(k)\mathbf{\Gamma}(k), \tag{7.41}$$

where the coherence matrix

$$\mathbf{\Gamma}(k) = \mathrm{diag}\left(|b_0(k)|, |b_1(k)|, |b_1(k)|, |b_1(k)|, |b_2(k)|, \ldots, |b_L(k)|\right)^{-2} \tag{7.42}$$

is an $N \times N$ diagonal matrix. In an anechoic environment, assuming plane-wave incidence from a direction Ω_s, the RTF vector $\mathbf{d}(k)$ is given by (7.37). By substituting (7.41) and (7.37) into (7.31), we find

$$\mathbf{w}_\mathrm{MVDR}(k) = \frac{Y_{00}(\Omega_\mathrm{s})\mathbf{\Gamma}^{-1}(k)\mathbf{y}^*(\Omega_\mathrm{s})}{\mathbf{y}^\mathrm{T}(\Omega_\mathrm{s})\mathbf{\Gamma}^{-1}(k)\mathbf{y}^*(\Omega_\mathrm{s})}. \tag{7.43}$$

Using $Y_{00}(\cdot) = 1/\sqrt{4\pi}$ and the spherical harmonic addition theorem [39], the beamformer weights simplify to

$$\mathbf{w}_\mathrm{MVDR}(k) = \frac{\sqrt{4\pi}\mathbf{\Gamma}^{-1}(k)\mathbf{y}^*(\Omega_\mathrm{s})}{\sum_{l=0}^{L} |b_l(k)|^2 (2l+1)}. \tag{7.44}$$

The beamformer output is given by

$$\begin{aligned} Z(k) &= \sqrt{4\pi}\frac{\sum_{l=0}^{L}\sum_{m=-l}^{l} |b_l(k)|^2\, Y_{lm}(\Omega_\mathrm{s})\widetilde{P}_{lm}(k)}{\sum_{l=0}^{L} |b_l(k)|^2 (2l+1)} \\ &= \frac{\sum_{l=0}^{L}\sum_{m=-l}^{l} b_l^*(k) Y_{lm}(\Omega_\mathrm{s}) P_{lm}(k)}{\sum_{l=0}^{L} |b_l(k)|^2 (2l+1)}. \end{aligned} \tag{7.45}$$

which is proportional to the output of a maximum WNG beamformer, as defined in (6.48), and therefore Property 7.2 holds.

7.3.4 Parametric Wiener Filter

The *parametric Wiener filter* (or *tradeoff filter*) balances noise reduction against speech distortion, and can be obtained by minimizing the cost function

$$J_{\mathrm{PWF},\mu}[\mathbf{w}(k)] = \mathrm{E}\left\{\left|\widetilde{X}_{\mathrm{f}}(k) - \widetilde{X}_{00}(k)\right|^2\right\} + \mu(k)\mathrm{E}\left\{\left|\widetilde{V}_{\mathrm{r}}(k)\right|^2\right\} \tag{7.46a}$$

$$= \phi_{\widetilde{X}_{00}}(k)\left|\mathbf{w}^{\mathrm{H}}(k)\mathbf{d}(k) - 1\right|^2 + \mu(k)\mathbf{w}^{\mathrm{H}}(k)\mathbf{\Phi}_{\widetilde{\mathbf{v}}}(k)\mathbf{w}(k). \tag{7.46b}$$

The first term on the right hand side of (7.46) is related to the speech distortion and the second term is equal to the residual noise power at the filter's output. The balance between speech distortion and noise reduction can be controlled using the parameter $\mu(k)$ ($\mu \geq 0$), which is referred to as a *tradeoff parameter*. It can readily be seen that for $\mu(k) = 1$, the cost function (7.46) is equal to the MSE cost function (7.21) used to derive the Wiener filter in Sect. 7.3.2.

For a given parameter $\mu(k)$, the parametric Wiener filter is then obtained using

$$\mathbf{w}_{\mathrm{PWF},\mu}(k) = \underset{\mathbf{w}(k)}{\arg\min}\, J_{\mathrm{PWF},\mu}[\mathbf{w}(k)]. \tag{7.47}$$

The filter that minimizes the cost function $J_{\mathrm{PWF},\mu}$ is computed as follows

$$\begin{aligned} \nabla_{\mathbf{w}^*_{\mathrm{PWF},\mu}} J_{\mathrm{PWF},\mu}\left[\mathbf{w}_{\mathrm{PWF},\mu}(k)\right] &= \mathbf{0}_N \\ \phi_{\widetilde{X}_{00}}(k)\left[\mathbf{d}(k)\mathbf{d}^{\mathrm{H}}(k)\mathbf{w}_{\mathrm{PWF},\mu}(k) - \mathbf{d}(k)\right] + \mu(k)\mathbf{\Phi}_{\widetilde{\mathbf{v}}}(k)\mathbf{w}_{\mathrm{PWF},\mu}(k) &= \mathbf{0}_N \\ \mathbf{\Phi}_{\widetilde{\mathbf{x}}}(k)\mathbf{w}_{\mathrm{PWF},\mu}(k) - \phi_{\widetilde{X}_{00}}(k)\mathbf{d}(k) + \mu(k)\mathbf{\Phi}_{\widetilde{\mathbf{v}}}(k)\mathbf{w}_{\mathrm{PWF},\mu}(k) &= \mathbf{0}_N. \end{aligned} \tag{7.48}$$

Finally, we obtain the parametric Wiener filter [16, 33]:

$$\mathbf{w}_{\mathrm{PWF},\mu}(k) = \phi_{\widetilde{X}_{00}}(k)\left[\mathbf{\Phi}_{\widetilde{\mathbf{x}}}(k) + \mu(k)\mathbf{\Phi}_{\widetilde{\mathbf{v}}}(k)\right]^{-1}\mathbf{d}(k). \tag{7.49}$$

Using the Woodbury matrix identity we can express (7.49) as

$$\mathbf{w}_{\mathrm{PWF},\mu}(k) = \frac{\phi_{\widetilde{X}_{00}}(k)\mathbf{\Phi}_{\widetilde{\mathbf{v}}}^{-1}(k)\mathbf{d}(k)}{\mu(k) + \phi_{\widetilde{X}_{00}}(k)\mathbf{d}^{\mathrm{H}}(k)\mathbf{\Phi}_{\widetilde{\mathbf{v}}}^{-1}(k)\mathbf{d}(k)} \tag{7.50a}$$

$$= \frac{\mathbf{\Phi}_{\widetilde{\mathbf{v}}}^{-1}(k)\mathbf{\Phi}_{\widetilde{\mathbf{p}}}(k) - \mathbf{I}_{N\times N}}{\mu(k) + \mathrm{tr}\left[\mathbf{\Phi}_{\widetilde{\mathbf{v}}}^{-1}(k)\mathbf{\Phi}_{\widetilde{\mathbf{p}}}(k)\right] - N}\mathbf{i}_N. \tag{7.50b}$$

The tradeoff parameter is often chosen in an ad-hoc way, keeping in mind that:

- $\mu(k) = 0$ corresponds to the MVDR filter (7.31);
- $\mu(k) = 1$ corresponds to the Wiener filter (7.25);
- $\mu > 1$ results in low residual noise at the expense of high speech distortion when compared to the Wiener filter [2]; and

- $\mu < 1$ results in high residual noise and low speech distortion compared to the Wiener filter [2].

We can see that the parametric Wiener filter is a special case of the maximum SNR filter in (7.19) for the case where the maximum SNR filter is computed using the scaling factor

$$\alpha(k) = \frac{\phi_{\widetilde{X}_{00}}(k)}{\mu(k) + \phi_{\widetilde{X}_{00}}(k)\mathbf{d}^{\mathrm{H}}(k)\mathbf{\Phi}_{\widetilde{\mathbf{v}}}^{-1}(k)\mathbf{d}(k)}. \tag{7.51}$$

7.3.5 Linearly Constrained Minimum Variance Filter

The MVDR filter in Sect. 7.3.3 seeks to suppress the noise signal $\widetilde{\mathbf{v}}(k)$ while maintaining a single distortionless constraint on the desired signal $\widetilde{X}_{00}(k)$. The LCMV filter is a generalization of the MVDR filter and is able to provide multiple distortionless constraints.

Let us now consider a scenario with $I \leq N$ directional sources such that

$$P(k, \mathbf{r}_q) = \sum_{\iota=1}^{I} H^{(\iota)}(k, \mathbf{r}_q)\, S^{(\iota)}(k) + V(k, \mathbf{r}_q) \tag{7.52a}$$

$$= \sum_{\iota=1}^{I} X^{(\iota)}(k, \mathbf{r}_q) + V(k, \mathbf{r}_q), \tag{7.52b}$$

where ι is the source index, $H^{(\iota)}(k, \mathbf{r}_q)$ is the acoustic transfer function from the ιth source to the microphone at position $\mathbf{r}_q$, $S^{(\iota)}(k)$ is the ιth source signal, and $X^{(\iota)}(k, \mathbf{r}_q)$ is the ιth received source signal. In the SHD, after mode strength compensation, we then have

$$\widetilde{P}_{lm}(k) = \sum_{\iota=1}^{I} \widetilde{X}_{lm}^{(\iota)}(k) + \widetilde{V}_{lm}(k) \tag{7.53a}$$

$$= \sum_{\iota=1}^{I} \mathbf{d}^{(\iota)}(k)\widetilde{X}_{00}^{(\iota)}(k) + \widetilde{V}_{lm}(k), \tag{7.53b}$$

where $\mathbf{d}^{(\iota)}(k)$ denotes the RTF of the ιth source. In vector notation, (7.53b) can be written for all eigenbeams as

$$\widetilde{\mathbf{p}}(k) = \mathbf{D}(k)\widetilde{\mathbf{x}}_{00}(k) + \widetilde{\mathbf{v}}(k), \tag{7.54}$$

where

$$\mathbf{D}(k) = \left[\mathbf{d}^{(1)}(k) \,\middle|\, \mathbf{d}^{(2)}(k) \middle| \ldots \middle| \mathbf{d}^{(I)}(k)\right] \tag{7.55}$$

is an $N \times I$ matrix, and $\widetilde{\mathbf{x}}_{00}(k) = [\widetilde{X}_{00}^{(1)}(k), \widetilde{X}_{00}^{(2)}(k), \ldots, \widetilde{X}_{00}^{(I)}(k)]^{\mathrm{T}}$ is a column vector of length I.

The desired signal that we wish to estimate can be written as

$$X_{\mathrm{d}}(k) = \sum_{\imath=1}^{I} Q^{(\imath)}(k)\, X_{00}^{(\imath)}(k), \tag{7.56}$$

where $Q^{(\imath)}(k)$ denotes the desired response for the $\imath$th source. When only the first source is desired, then $Q^{(1)}(k) = 1$ and $Q^{(\imath)}(k) = 0$ for $\imath \in \{2, 3, \ldots, I\}$. Alternatively, when $Q^{(\imath)}(k) = 1$ for all $\imath$, all directional sources are extracted.

The LCMV filter can be used to obtain an estimate of the desired signal $X_{\mathrm{d}}(k)$ by minimizing the residual noise at the output of the beamformer, given by $\mathrm{E}\{|\mathbf{w}^{\mathrm{H}}(k)\mathbf{v}(k)|^2\}$, subject to the constraint

$$\mathbf{w}^{\mathrm{H}}(k)\mathbf{D}(k) = \mathbf{q}^{\mathrm{T}}(k), \tag{7.57}$$

where $\mathbf{q}(k) = [Q^{(1)}(k), Q^{(2)}(k), \ldots, Q^{(I)}(k)]^{\mathrm{T}}$. Mathematically, this problem can be formulated as

$$\mathbf{w}_{\mathrm{LCMV}}(k) = \arg\min_{\mathbf{w}(k)} \mathbf{w}^{\mathrm{H}}(k)\mathbf{\Phi}_{\widetilde{\mathbf{v}}}(k)\mathbf{w}(k) \quad \text{subject to} \quad \mathbf{w}^{\mathrm{H}}(k)\mathbf{D}(k) = \mathbf{q}^{\mathrm{T}}(k). \tag{7.58}$$

The solution is the well-known LCMV filter given by

$$\mathbf{w}_{\mathrm{LCMV}}(k) = \mathbf{\Phi}_{\widetilde{\mathbf{v}}}^{-1}(k)\mathbf{D}(k)\left[\mathbf{D}^{\mathrm{H}}(k)\mathbf{\Phi}_{\widetilde{\mathbf{v}}}^{-1}(k)\mathbf{D}(k)\right]^{-1}\mathbf{q}^{*}(k). \tag{7.59}$$

For this solution to exist, the noise PSD matrix $\mathbf{\Phi}_{\widetilde{\mathbf{v}}}$ needs to have full rank and the columns of the matrix $\mathbf{D}(k)$ need to be linearly independent [36].

The LCMV filter can be interpreted as a two stage spatial processor that first computes I signals given by $\mathbf{D}^{\mathrm{H}}(k)\mathbf{\Phi}_{\widetilde{\mathbf{v}}}^{-1}(k)\widetilde{\mathbf{p}}(k)$. These signals are then combined using $\mathbf{q}^{\mathrm{T}}\left[\mathbf{D}^{\mathrm{H}}(k)\mathbf{\Phi}_{\widetilde{\mathbf{v}}}^{-1}(k)\mathbf{D}(k)\right]^{-1}$ to compute the output signal $Z(k)$ of the LCMV filter.

Instead of minimizing the noise power $\phi_{\widetilde{V}_{\mathrm{r}}}$ at the output of the beamformer, one can minimize the total power of the beamformer's output (i.e., desired speech plus residual noise). This optimization problem can be written as

$$\min_{\mathbf{w}(k)} \mathbf{w}^{\mathrm{H}}(k)\mathbf{\Phi}_{\widetilde{\mathbf{p}}}(k)\mathbf{w}(k) \quad \text{subject to} \quad \mathbf{w}^{\mathrm{H}}(k)\mathbf{D}(k) = \mathbf{q}^{\mathrm{T}}(k), \tag{7.60}$$

and its solution is the linearly constrained minimum power (LCMP) filter

$$\mathbf{w}_{\mathrm{LCMP}}(k) = \mathbf{\Phi}_{\widetilde{\mathbf{p}}}^{-1}(k)\mathbf{D}(k)\left[\mathbf{D}^{\mathrm{H}}(k)\mathbf{\Phi}_{\widetilde{\mathbf{p}}}^{-1}(k)\mathbf{D}(k)\right]^{-1}\mathbf{q}^{*}(k). \tag{7.61}$$

As we do not have access to the true RTF vectors $\mathbf{D}(k)$, we need an estimate of $\mathbf{D}(k)$ to compute the LCMV and LCMP filters. It can be shown that the LCMV and LCMP filters are equivalent only when the estimated and true RTF vectors are equal. The LCMP filter is, however, known to be sensitive to RTF estimation errors [36].

7.3.6 Generalized Sidelobe Canceller Structure

Although in the previous sections we presented closed-form solutions to various constrained optimization problems, such as the MPDR and LCMP, in the past it was also common to find the solution using adaptive algorithms. A major advantage of using adaptive algorithms is that they do not require an estimate of the noise PSD matrix. However, they can become cumbersome when dealing with constrained optimization problems. The generalized sidelobe canceller (GSC), proposed by Griffiths and Jim [14], is a filter structure that allows any of the previously considered constrained optimization problems to be formulated as unconstrained optimization problems, which significantly simplifies the adaptive algorithms. The MVDR, MPDR, LCMV, and LCMP filters can all be implemented using the GSC structure. The performance of the GSC was analyzed by several researchers (e.g. [6, 11, 29]), and an extension of the GSC taking into account the acoustic transfer functions (ATFs) was first proposed by Gannot et al. in [12].

The weights of the filters corresponding to one of the constrained optimization problems formulated in this chapter span an $N = (L+1)^2$ dimensional space that can be divided into two orthogonal subspaces: a constraint subspace and an orthogonal subspace. The constraint subspace with rank I is defined by the column space $\mathbf{D}(k)$ defined in (7.55), and the orthogonal subspace with rank $N - I$ is defined by the left null space of $\mathbf{D}(k)$. From this perspective, an alternative representation of, for example, (7.59) is obtained by decomposing $\mathbf{w}_{\mathrm{LCMV}}(k)$ into one component $\mathbf{w}_{\mathrm{c}}(k)$ that lies in the space spanned by $\mathbf{D}(k)$, and another component $\mathbf{w}_{\mathrm{c}}^{\perp}(k)$ that lies in the left null space of $\mathbf{D}(k)$, i.e.,

$$\mathbf{w}_{\mathrm{LCMV}}(k) = \mathbf{w}_{\mathrm{c}}(k) + \mathbf{w}_{\mathrm{c}}^{\perp}(k). \tag{7.62}$$

In Fig. 7.2 the constrained minimization is illustrated for two eigenbeams and a single directional source (i.e., $I = 1$). The axes represent the two filter weights. The dashed contour lines represent the power of the residual noise (i.e., $\mathbf{w}^{\mathrm{H}}(k)\boldsymbol{\Phi}_{\tilde{\mathbf{v}}}(k)\mathbf{w}(k)$), and the straight line represents the solutions for which the constraint $\mathbf{w}^{\mathrm{H}}(k)\mathbf{d}(k) = 1$ is satisfied. The vector $\mathbf{w}_{\mathrm{MVDR}}(k)$ represents the MVDR filter that minimizes the residual noise and satisfies the constraint. The weights of the filter $\mathbf{w}_{\mathrm{c}}(k)$ satisfy the constraint. Other than when $\boldsymbol{\Phi}_{\tilde{\mathbf{v}}}(k)$ is a scaled identity matrix, these solutions do not minimize the residual noise power. Figure 7.3 illustrates the structure of (7.62). The filter $\mathbf{w}_{\mathrm{c}}(k)$ is part of the upper branch, while the filter $\mathbf{w}_{\mathrm{c}}^{\perp}(k)$ (not shown in Fig. 7.2) is part of the lower branch.

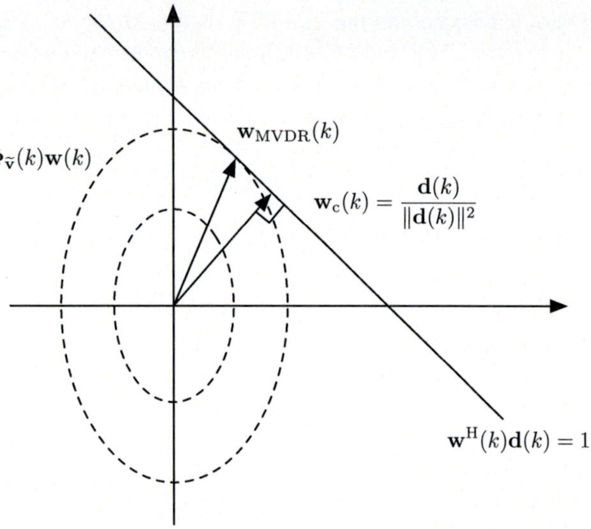

Fig. 7.2 Illustration of the constrained minimization for a single directional source (i.e., $I = 1$)

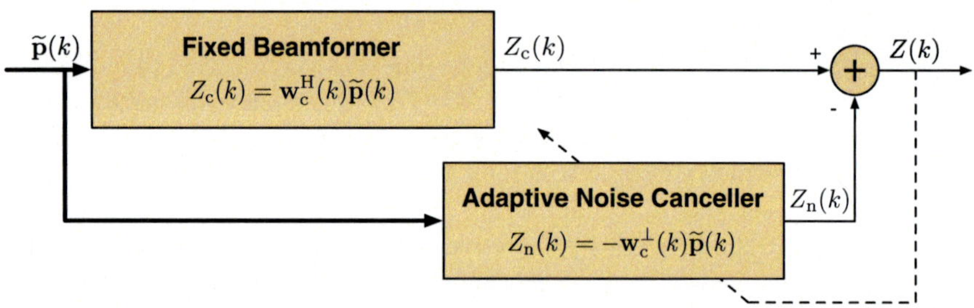

Fig. 7.3 Block diagram of a signal-dependent beamformer implemented using the structure of (7.62)

In the context of the GSC, the filter $\mathbf{w}_{\mathrm{c}}(k)$ is also known as the *quiescent filter*, and can be obtained by projecting the LCMV filter on the constraint subspace spanned by the RTFs. Since the RTFs are not necessarily orthonormal, a suitable projection matrix of size $N \times N$ is given by

$$\mathbf{P}_{\mathrm{c}}(k) = \mathbf{D}(k)\left[\mathbf{D}^{\mathrm{H}}(k)\mathbf{D}(k)\right]^{-1}\mathbf{D}^{\mathrm{H}}(k), \tag{7.63}$$

such that

$$\mathbf{w}_c(k) = \mathbf{P}_c(k)\,\mathbf{w}_{\mathrm{LCMV}}(k) \tag{7.64a}$$
$$= \mathbf{D}(k)\left[\mathbf{D}^{\mathrm{H}}(k)\mathbf{D}(k)\right]^{-1}\mathbf{D}^{\mathrm{H}}(k)\,\mathbf{w}_{\mathrm{LCMV}}(k) \tag{7.64b}$$
$$= \mathbf{D}(k)\left[\mathbf{D}^{\mathrm{H}}(k)\mathbf{D}(k)\right]^{-1}\mathbf{q}^*(k). \tag{7.64c}$$

The filter $\mathbf{w}_c^{\perp}(k)$ can be obtained by projecting the LCMV filter on the orthogonal subspace using the projection matrix

$$\mathbf{P}_o(k) = \mathbf{C}_n(k)\left[\mathbf{C}_n^{\mathrm{H}}(k)\mathbf{C}_n(k)\right]^{-1}\mathbf{C}_n^{\mathrm{H}}(k), \tag{7.65}$$

where $\mathbf{C}_n(k)$ is a matrix of size $N \times (N - I')$ with $0 \le I' \le I$ that is chosen such that

$$\mathbf{D}^{\mathrm{H}}(k)\mathbf{C}_n(k) = \mathbf{0}_{I\times(N-I')}. \tag{7.66}$$

The rank of the matrix $\mathbf{C}_n(k)$ is equal to $N - I$, i.e., the rank of the orthogonal subspace. For $I' < I$, the columns of $\mathbf{C}_n(k)$ form an over-complete basis of the orthogonal subspace. It should be noted that any vector that lies in the column space of $\mathbf{C}_n(k)$, and hence in the left null space of $\mathbf{D}(k)$, lies in the orthogonal subspace. The filter $\mathbf{w}_c^{\perp}(k)$ is then given by [36, Sect. 6.7.3]

$$\mathbf{w}_c^{\perp}(k) = \mathbf{P}_o(k)\,\mathbf{w}_{\mathrm{LCMV}}(k) \tag{7.67a}$$
$$= -\mathbf{C}_n(k)\left[\mathbf{C}_n^{\mathrm{H}}(k)\boldsymbol{\Phi}_{\widetilde{\mathbf{v}}}(k)\mathbf{C}_n(k)\right]^{-1}\mathbf{C}_n^{\mathrm{H}}(k)\boldsymbol{\Phi}_{\widetilde{\mathbf{v}}}(k)\mathbf{w}_c(k). \tag{7.67b}$$

In the context of the GSC, the matrix $\mathbf{C}_n(k)$ is known as the *blocking matrix*. The signals at the output of this blocking matrix are referred to as *reference signals* and are given by

$$\widetilde{\mathbf{u}}(k) = \mathbf{C}_n^{\mathrm{H}}(k)\widetilde{\mathbf{p}}(k). \tag{7.68}$$

The number of reference signals depends on the dimensions of the blocking matrix. If the blocking matrix is correctly constructed, then the reference signals are uncorrelated with the source signals $S^{(1)}(k), S^{(2)}(k), \ldots, S^{(I)}(k)$.

By substituting (7.67b) into (7.62) we can represent the LCMV filter as

$$\mathbf{w}_{\mathrm{LCMV}}(k) = \mathbf{w}_c(k) - \mathbf{C}_n(k)\mathbf{w}_n(k), \tag{7.69}$$

where

$$\mathbf{w}_n(k) = [\mathbf{C}_n^{\mathrm{H}}(k)\boldsymbol{\Phi}_{\widetilde{\mathbf{v}}}(k)\mathbf{C}_n(k)]^{-1}\mathbf{C}_n^{\mathrm{H}}(k)\boldsymbol{\Phi}_{\widetilde{\mathbf{v}}}(k)\mathbf{w}_c(k) \tag{7.70}$$

is known as the *noise cancellation filter*. The structure of (7.69) is more commonly known as the GSC structure and is illustrated in Fig. 7.4.

Following the above derivation, it is evident that the GSC is another implementation of the LCMV filter. However, when inspecting the filters, this equivalence might

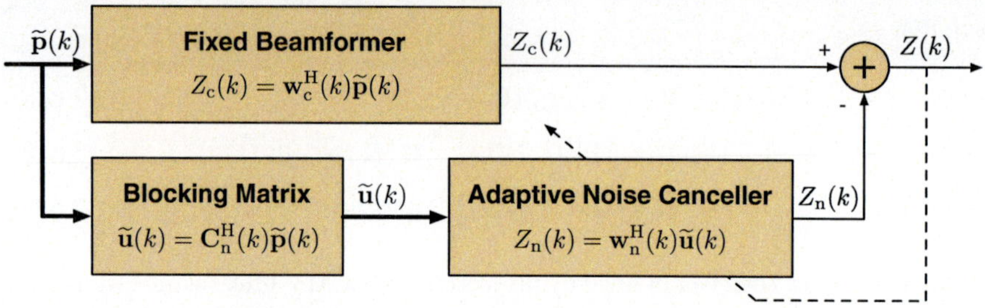

Fig. 7.4 Block diagram of a signal-dependent beamformer implemented using the GSC structure of (7.69)

not be that evident. In [9], Breed and Strauss showed using a short and elegant proof that both implementations are equivalent.

Finally, we must determine the blocking matrix $\mathbf{C}_n(k)$ that satisfies the condition in (7.66). As a matter of fact, there are infinitely many blocking matrices that satisfy this condition. A frequently used blocking matrix is known as the *sparse blocking matrix* [12]. For $I = 1$, if we define $\tilde{X}_{00}(k)$ as the desired source signal, an example of the sparse blocking matrix is given by

$$\mathbf{C}_n(k) = \begin{pmatrix} -D^*_{1(-1)}(k) & -D^*_{10}(k) & \cdots & -D^*_{LL}(k) \\ 1 & 0 & 0 & 0 \\ 0 & 1 & 0 & 0 \\ 0 & 0 & \ddots & 0 \\ 0 & 0 & 0 & 1 \end{pmatrix} \tag{7.71}$$

for which $\mathbf{D}^H(k)\mathbf{C}_n(k) = \left(\mathbf{d}^{(1)}(k)\right)^H \mathbf{C}_n(k) = \mathbf{0}_{1\times N-1}$. An extension of the sparse blocking matrix for multiple constraints was presented in [27]. Another popular blocking matrix is the *eigen-space blocking matrix* that is given by

$$\mathbf{C}_n(k) = \mathbf{I}_{N\times N} - \mathbf{D}(k)\left[\mathbf{D}^H(k)\mathbf{D}(k)\right]^{-1}\mathbf{D}^H(k). \tag{7.72}$$

The signal leakage and the blocking ability of the sparse blocking matrix and of the eigen-space blocking matrix were analyzed and compared in [27]. It was analytically proven that the blocking abilities of both blocking matrices are equivalent, provided that the estimate of $\mathbf{D}(k)$, which in practice is required to compute $\mathbf{C}_n(k)$, corresponds to the true propagation vector. The dimensions of the sparse blocking matrix are $N \times (N - I)$, while the dimensions of the eigen-space blocking matrix are $N \times N$, therefore the sparse blocking matrix can be smaller than the eigen-space blocking matrix. Consequently, the length of the noise cancellation filter is also smaller when employing the sparse blocking matrix. Hence, the overall complexity of the GSC is smaller when using the sparse rather than the eigen-space blocking matrix.

It is interesting to note that the noise cancellation filter in (7.70) is a multichannel Wiener filter [5, 29]. Therefore, the filter can be obtained by minimizing

$$\mathrm{E}\left\{\left|\mathbf{w}_{\mathrm{c}}^{\mathrm{H}}(k)\widetilde{\mathbf{v}}(k) - \mathbf{w}_{\mathrm{n}}^{\mathrm{H}}(k)\mathbf{C}_{\mathrm{n}}^{\mathrm{H}}(k)\widetilde{\mathbf{v}}(k)\right|^2\right\}. \tag{7.73}$$

While the closed-form solution is given by (7.70), a solution can also be obtained by minimizing (7.73) adaptively. For example, the filter update equation corresponding to a normalized least mean squares algorithm is given by [12]

$$\mathbf{w}_{\mathrm{n}}(\ell+1,k) = \begin{cases} \mathbf{w}_{\mathrm{n}}(\ell,k) - \frac{\vartheta_{\mathrm{s}}}{\mathrm{tr}\{\boldsymbol{\Phi}_{\widetilde{\mathbf{u}}}(\ell,k)\}}\widetilde{\mathbf{u}}(\ell,k)Z^*(\ell,k) & \text{if desired sources inactive} \\ \mathbf{w}_{\mathrm{n}}(\ell,k) & \text{otherwise} \end{cases} \tag{7.74}$$

where $\boldsymbol{\Phi}_{\widetilde{\mathbf{u}}}(\ell,k) = \mathrm{E}\left\{\widetilde{\mathbf{u}}(\ell,k)\widetilde{\mathbf{u}}^{\mathrm{H}}(\ell,k)\right\}$, ℓ is the time frame index, ϑ_{s} is the step size, and

$$Z(\ell,k) = [\mathbf{w}_{\mathrm{c}}(k) - \mathbf{C}_{\mathrm{n}}(k)\mathbf{w}_{\mathrm{n}}(\ell,k)]^{\mathrm{H}}\,\widetilde{\mathbf{p}}(\ell,k). \tag{7.75}$$

7.4 Relative Transfer Function Estimation

The weights of many of the previously derived signal-dependent beamformers are expressed in terms of the RTF vector $\mathbf{d}(k)$. In the context of SHD processing, the RTF describes the linear relationship between the desired signal vector $\widetilde{\mathbf{x}}(k)$ and a reference signal. In Sect. 7.1, we used the received source component $\widetilde{X}_{00}$ of the signal received at $\mathcal{M}_{\mathrm{ref}}$ as a reference signal, such that

$$\widetilde{\mathbf{x}}(k) = \mathbf{d}(k)\widetilde{X}_{00}(k). \tag{7.76}$$

It is important to note that the choice of the reference signal defines the desired signal that the beamformer seeks to estimate. As a matter of fact, one could instead use another eigenbeam, or even a linear combinations of eigenbeams as a reference signal.

In an anechoic environment, assuming plane-wave incidence from a direction Ω_{s}, the RTF vector $\mathbf{d}(k)$ is given by [8]

$$\mathbf{d}(k) = \frac{\mathbf{y}^*(\Omega_{\mathrm{s}})}{Y_{00}^*(\Omega_{\mathrm{s}})}, \tag{7.77}$$

where Ω_{s} denotes the direction of arrival (DOA) of the desired direct sound.

It should be noted that in a reverberant environment, and when the shorttime Fourier transform (STFT) frame length is sufficiently long such that the *multiplicative transfer function* approximation [1] holds, $\widetilde{X}_{00}(k)$ contains the direct sound as well as early reflections and reverberation. The RTF vector, which is sometimes also referred to as the spatial prediction vector, thus depends on the position of the source and the spherical microphone array as well as the room characteristics.

Since the performance of signal-dependent beamformers in reverberant environments strongly depends on the accuracy of the estimated RTF vector, many different estimators have been developed. In the following subsections, we present some of the most frequently used estimators, namely the covariance subtraction method [2], the generalized eigenvector method [25], and the temporal averaging method. The first two methods require an estimate of the observed signal PSD matrix and the noise PSD matrix. A theoretical comparison between the covariance subtraction method and the so-called covariance whitening method that is closely related to the generalized eigenvector method can be found in [26]. The third method, the temporal averaging method, exploits the fact that the statistics of the desired signal change more rapidly compared to the statistics of the noise, and only requires an estimate of the observed signal PSD matrix.

7.4.1 Covariance Subtraction Method

The most straightforward approach is to estimate the RTF vector in the MSE sense, i.e.,

$$\widehat{\mathbf{d}}(k) = \arg\min_{\mathbf{d}(k)} \mathrm{E}\left\{\left\|\widetilde{\mathbf{x}}(k) - \mathbf{d}(k)\widetilde{X}_{00}(k)\right\|^2\right\}. \tag{7.78}$$

The solution is given by [2]

$$\begin{aligned}\widehat{\mathbf{d}}(k) &= \frac{\mathrm{E}\left\{\widetilde{\mathbf{x}}(k)\widetilde{X}_{00}^*(k)\right\}}{\mathrm{E}\left\{\left|\widetilde{X}_{00}(k)\right|^2\right\}} \\ &= \frac{\mathbf{\Phi}_{\widetilde{\mathbf{x}}}(k)\mathbf{i}_N}{\mathbf{i}_N^{\mathrm{T}}\mathbf{\Phi}_{\widetilde{\mathbf{x}}}(k)\mathbf{i}_N},\end{aligned} \tag{7.79}$$

where $\mathbf{i}_N = [1\ 0\ 0\ \dots\ 0]^{\mathrm{T}}$ is a column vector of length N.

Since the PSD matrix $\mathbf{\Phi}_{\widetilde{\mathbf{x}}}(k)$ is unobservable in practice, it is commonly expressed in terms of the PSD matrix of the noise $\mathbf{\Phi}_{\widetilde{\mathbf{v}}}(k)$ and the PSD matrix of the observed signals $\mathbf{\Phi}_{\widetilde{\mathbf{p}}}(k)$. Using (7.8) we can rewrite (7.79) as

$$\widehat{\mathbf{d}}_{\mathrm{CV}}(k) = \frac{\left[\mathbf{\Phi}_{\widetilde{\mathbf{p}}}(k) - \mathbf{\Phi}_{\widetilde{\mathbf{v}}}(k)\right]\mathbf{i}_N}{\mathbf{i}_N^{\mathrm{T}}\left[\mathbf{\Phi}_{\widetilde{\mathbf{p}}}(k) - \mathbf{\Phi}_{\widetilde{\mathbf{v}}}(k)\right]\mathbf{i}_N}. \tag{7.80}$$

7.4.2 Generalized Eigenvector Method

The generalized eigenvector method proposed in [25] makes use of the fact that the PSD matrix $\mathbf{\Phi}_{\widetilde{\mathbf{x}}}(k)$ is rank-one (i.e., only one desired source is present). If $\mathbf{d}(k)$

denotes the RTF of the desired source then the PSD matrix of the observed signal after mode strength compensation can be expressed as

$$\mathbf{\Phi}_{\widetilde{\mathbf{p}}}(k) = \phi_{\widetilde{X}_{00}}(k)\mathbf{d}(k)\mathbf{d}^{\mathrm{H}}(k) + \mathbf{\Phi}_{\widetilde{\mathbf{v}}}(k). \tag{7.81}$$

The generalized eigenvalue decomposition of the matrix pencil $(\mathbf{\Phi}_{\widetilde{\mathbf{p}}}, \mathbf{\Phi}_{\widetilde{\mathbf{v}}})$ can be written as

$$\mathbf{\Phi}_{\widetilde{\mathbf{p}}}\mathbf{u} = \lambda\mathbf{\Phi}_{\widetilde{\mathbf{v}}}\mathbf{u} \tag{7.82a}$$

$$\left(\phi_{\widetilde{X}_{00}}\mathbf{d}\mathbf{d}^{\mathrm{H}} + \mathbf{\Phi}_{\widetilde{\mathbf{v}}}\right)\mathbf{u} = \lambda\mathbf{\Phi}_{\widetilde{\mathbf{v}}}\mathbf{u} \tag{7.82b}$$

$$\left(\phi_{\widetilde{X}_{00}}\mathbf{d}\mathbf{d}^{\mathrm{H}}\right)\mathbf{u} = (\lambda - 1)\mathbf{\Phi}_{\widetilde{\mathbf{v}}}\mathbf{u}, \tag{7.82c}$$

where λ and $\mathbf{u}$ denote an eigenvalue and corresponding eigenvector. The eigenvectors corresponding to the eigenvalues with values other than one span the subspace of the desired signal. Since there is only one desired source, there is only one eigenvalue that is larger than one. It can be shown that the corresponding eigenvector, denoted by $\mathbf{u}_{\mathrm{max}}$, is equal to a scaled version of the desired source ATF [25].

Solving for $\mathbf{d}(k)$ leads to

$$\mathbf{d}(k) = \underbrace{\frac{\lambda - 1}{\phi_{\widetilde{X}_{00}}(k)\mathbf{d}^{\mathrm{H}}(k)\mathbf{u}_{\mathrm{max}}(k)}}_{\text{scalar}}\mathbf{\Phi}_{\widetilde{\mathbf{v}}}(k)\mathbf{u}_{\mathrm{max}}(k). \tag{7.83}$$

Hence, the RTF vector is a scaled and rotated version of the eigenvector $\mathbf{u}_{\mathrm{max}}(k)$.

As the first element of $\mathbf{d}(k)$ is by definition equal to one, there is no need to compute the scalar and the RTF can be obtained directly by dividing $\mathbf{\Phi}_{\widetilde{\mathbf{v}}}(k)\mathbf{u}_{\mathrm{max}}$ by its first element, i.e.,

$$\mathbf{d}(k) = \frac{\mathbf{\Phi}_{\widetilde{\mathbf{v}}}(k)\mathbf{u}_{\mathrm{max}}(k)}{\mathbf{i}_N^{\mathrm{T}}\ \mathbf{\Phi}_{\widetilde{\mathbf{v}}}(k)\mathbf{u}_{\mathrm{max}}(k)\ \mathbf{i}_N}. \tag{7.84}$$

As mentioned in Sect. 7.3.1, the vector $\mathbf{u}_{\mathrm{max}}(k)$ is also a maximum SNR filter. By substituting $\mathbf{u}_{\mathrm{max}}(k)$ with $\mathbf{w}_{\mathrm{max}}(k) = \alpha(k)\mathbf{\Phi}_{\widetilde{\mathbf{v}}}^{-1}(k)\mathbf{d}(k)$ in (7.84), it follows that the right-hand-side is equal to $\mathbf{d}(k)$.

7.4.3 *Temporal Averaging Method*

When the desired signal is speech, we can assume that the statistics of the noise vary slowly compared to the statistics of the desired signal [12]. The observed signal can be expressed as

$$\widetilde{P}_{lm}(\ell, k) = D_{lm}(k)\widetilde{P}_{00}(\ell, k) + \widetilde{U}_{lm}(\ell, k), \tag{7.85}$$

where ℓ denotes the time frame index and

$$\widetilde{U}_{lm}(\ell,k) = \widetilde{V}_{lm}(\ell,k) - D_{lm}(k)\widetilde{V}_{00}(\ell,k). \quad (7.86)$$

Multiplying both sides of (7.85) by $\widetilde{P}^*_{00}(\ell,k)$ and taking the expectation yields:

$$\widehat{\phi}_{\widetilde{P}_{lm}\widetilde{P}_{00}}(\ell,k) = D_{lm}(k)\widehat{\phi}_{\widetilde{P}_{00}}(\ell,k) + \phi_{\widetilde{U}_{lm}\widetilde{P}_{00}}(\ell,k) + \epsilon_{lm}(\ell,k),$$

with

$$\epsilon_{lm}(\ell,k) = \widehat{\phi}_{\widetilde{U}_{lm}\widetilde{P}_{00}}(\ell,k) - \phi_{\widetilde{U}_{lm}\widetilde{P}_{00}}(\ell,k), \quad (7.87)$$

where $\widehat{\phi}_{\widetilde{P}_{lm}\widetilde{P}_{00}}(\ell,k)$, $\widehat{\phi}_{\widetilde{P}_{00}}(\ell,k)$, and $\widehat{\phi}_{\widetilde{U}_{lm}\widetilde{P}_{00}}(\ell,k)$ are respectively estimates of

$$\begin{aligned}
\phi_{\widetilde{P}_{lm}\widetilde{P}_{00}}(\ell,k) &= \mathrm{E}\left\{\widetilde{P}_{lm}(\ell,k)\widetilde{P}^*_{00}(\ell,k)\right\}, \\
\phi_{\widetilde{P}_{00}}(\ell,k) &= \mathrm{E}\left\{|\widetilde{P}_{00}(\ell,k)|^2\right\}, \text{ and} \\
\phi_{\widetilde{U}_{lm}\widetilde{P}_{00}}(\ell,k) &= \mathrm{E}\left\{\widetilde{U}_{lm}(\ell,k)\widetilde{P}^*_{00}(\ell,k)\right\}.
\end{aligned}$$

Within a short time period of T time frames we can assume the noise is stationary such that

$$\phi_{\widetilde{U}_{lm}\widetilde{P}_{00}}(\ell,k) = \phi_{\widetilde{U}_{lm}\widetilde{P}_{00}}(k). \quad (7.88)$$

We can then combine the information obtained using T time frames and construct the following overdetermined set of equations:

$$\begin{bmatrix} \widehat{\phi}_{\widetilde{P}_{lm}\widetilde{P}_{00}}(\ell,k) \\ \widehat{\phi}_{\widetilde{P}_{lm}\widetilde{P}_{00}}(\ell-1,k) \\ \vdots \\ \widehat{\phi}_{\widetilde{P}_{lm}\widetilde{P}_{00}}(\ell-T+1,k) \end{bmatrix} = \begin{bmatrix} \widehat{\phi}_{\widetilde{P}_{00}}(\ell,k) & 1 \\ \widehat{\phi}_{\widetilde{P}_{00}}(\ell-1,k) & 1 \\ \vdots & \\ \widehat{\phi}_{\widetilde{P}_{00}}(\ell-T+1,k) & 1 \end{bmatrix} \begin{bmatrix} D_{lm}(k) \\ \phi_{\widetilde{U}_{lm}\widetilde{P}_{00}}(k) \end{bmatrix} + \begin{bmatrix} \epsilon_{lm}(\ell,k) \\ \epsilon_{lm}(\ell-1,k) \\ \vdots \\ \epsilon_{lm}(\ell-T+1,k) \end{bmatrix}.$$

An unbiased estimate of $D_{lm}(k)$ at time frame ℓ can then be computed as [32]

$$\widehat{D}_{lm}(\ell,k) = \frac{\left\langle \widehat{\phi}_{\widetilde{P}_{00}}(\ell,k)\widehat{\phi}_{\widetilde{P}_{lm}\widetilde{P}_{00}}(\ell,k)\right\rangle - \left\langle \widehat{\phi}_{\widetilde{P}_{00}}(\ell,k)\right\rangle\left\langle \widehat{\phi}_{\widetilde{P}_{lm}\widetilde{P}_{00}}(\ell,k)\right\rangle}{\left\langle \widehat{\phi}^2_{\widetilde{P}_{00}}(\ell,k)\right\rangle - \left\langle \widehat{\phi}_{\widetilde{P}_{00}}(\ell,k)\right\rangle^2}$$

where $\langle\cdot\rangle$ denotes a time averaging operator defined as

$$\langle A(\ell,k)\rangle \triangleq \frac{1}{T}\sum_{\ell'=0}^{T-1} A(\ell-\ell',k).$$

7.5 Chapter Summary

Signal-dependent beamformers adaptively achieve optimal performance in terms of signal-dependent performance measures, such as the speech distortion index, noise reduction factor or MSE. This chapter derived a number of beamformers based on these measures: the maximum SNR filter, the Wiener filter, the MVDR filter, the parametric Wiener filter and the LCMV filter. The weights of all these beamformers depend on the second-order statistics of the desired and noise signals. Methods for estimating these statistics remain a challenge and topic of active research; some possible approaches will be explored in Chap. 9. Finally, it was shown that for certain specific noise fields, the MVDR beamformer is equivalent to the signal-independent maximum directivity and maximum WNG beamformers introduced in Chap. 6.

References

1. Avargel, Y., Cohen, I.: On multiplicative transfer function approximation in the short-time Fourier transform domain. IEEE Signal Process. Lett. **14**(5), 337–340 (2007). doi:10.1109/LSP.2006.888292
2. Benesty, J., Chen, J., Habets, E.A.P.: Speech Enhancement in the STFT Domain. Springer Briefs in Electrical and Computer Engineering. Springer, Heidelberg (2011)
3. Benesty, J., Chen, J., Huang, Y.: Microphone Array Signal Processing. Springer, Berlin (2008)
4. Benesty, J., Makino, S., Chen, J. (eds.): Speech Enhancement. Springer, Heidelberg (2005)
5. Bitzer, J., Kammeyer, K.D., Simmer, K.U.: An alternative implementation of the superdirective beamformer. In: Proceedings of the IEEE Workshop on Applications of Signal Processing to Audio and Acoustics. New Paltz, New York (1999)
6. Bitzer, J., Simmer, K., Kammeyer, K.D.: Theoretical noise reduction limits of the generalized sidelobe canceller for speech enhancement. In: Proceedings of the IEEE International Conference on Acoustics, Speech and Signal Processing (ICASSP), vol. 5, pp. 2965–2968 (1999)
7. Brandwood, D.H.: A complex gradient operator and its application in adaptive array theory. In: Proceedings of the IEEE **130**(1, Parts F and H), 11–16 (1983)
8. Braun, S., Jarrett, D.P., Fischer, J., Habets, E.A.P.: An informed spatial filter for dereverberation in the spherical harmonic domain. In: Proceedings of the IEEE International Conference on Acoustics, Speech and Signal Processing (ICASSP), pp. 669–673. Vancouver, Canada (2013)
9. Breed, B.R., Strauss, J.: A short proof of the equivalence of LCMV and GSC beamforming. IEEE Signal Process. Lett. **9**(6), 168–169 (2002). doi:10.1109/LSP.2002.800506
10. Capon, J.: High resolution frequency-wavenumber spectrum analysis. Proc. IEEE **57**, 1408–1418 (1969)
11. Cohen, I.: Analysis of two-channel generalized sidelobe canceller with post-filtering. IEEE Trans. Speech Audio Process. **11**(6), 684–699 (2003)
12. Gannot, S., Burshtein, D., Weinstein, E.: Signal enhancement using beamforming and nonstationarity with applications to speech. IEEE Trans. Signal Process. **49**(8), 1614–1626 (2001)
13. Gannot, S., Cohen, I.: Adaptive beamforming and postfiltering. In: Benesty, J., Sondhi, M.M., Huang, Y. (eds.) Springer Handbook of Speech Processing, Chap. 47. Springer, Heidelberg (2008)
14. Griffiths, L.J., Jim, C.W.: An alternative approach to linearly constrained adaptive beamforming. IEEE Trans. Antennas Propag. **30**(1), 27–34 (1982)
15. Habets, E.A.P., Benesty, J., Cohen, I., Gannot, S., Dmochowski, J.: New insights into the MVDR beamformer in room acoustics. IEEE Trans. Audio, Speech, Lang. Process. **18**, 158–170 (2010)

16. Habets, E.A.P., Benesty, J., Gannot, S., Cohen, I.: The MVDR beamformer for speech enhancement. In: Cohen, I., Benesty, J., Gannot, S. (eds.) Speech Processing in Modern Communication: Challenges and Perspectives, Chap. 9. Springer, Heidelberg (2010)
17. Habets, E.A.P., Benesty, J., Naylor, P.A.: A speech distortion and interference rejection constraint beamformer. IEEE Trans. Audio, Speech, Lang. Process. **20**(3), 854–867 (2012)
18. ITU-T: Objective Measurement of Active Speech Level (1993)
19. Chen, J., Benesty, Y.H., Doclo, S.: New insights into the noise reduction Wiener filter. IEEE Trans. Audio, Speech, Lang. Process. **14**, 1218–1234 (2006)
20. Jarrett, D.P., Habets, E.A.P.: On the noise reduction performance of a spherical harmonic domain tradeoff beamformer. IEEE Signal Process. Lett. **19**(11), 773–776 (2012)
21. Jarrett, D.P., Habets, E.A.P., Benesty, J., Naylor, P.A.: A tradeoff beamformer for noise reduction in the spherical harmonic domain. In: Proceedings of the International Workshop Acoustics, Signal Enhancement (IWAENC). Aachen, Germany (2012)
22. Jarrett, D.P., Habets, E.A.P., Naylor, P.A.: Spherical harmonic domain noise reduction using an MVDR beamformer and DOA-based second-order statistics estimation. In: Proceedings of the IEEE International Conference on Acoustics, Speech and Signal Processing (ICASSP), pp. 654–658. Vancouver, Canada (2013)
23. Jarrett, D.P., Thiergart, O., Habets, E.A.P., Naylor, P.A.: Coherence-based diffuseness estimation in the spherical harmonic domain. In: Proceedings of the IEEE Convention of Electrical and Electronics Engineers in Israel (IEEEI). Eilat, Israel (2012)
24. Kuttruff, H.: Room Acoustics, 4th edn. Taylor and Francis, London (2000)
25. Markovich, S., Gannot, S., Cohen, I.: Multichannel eigenspace beamforming in a reverberant noisy environment with multiple interfering speech signals. IEEE Trans. Audio, Speech, Lang. Process. **17**(6), 1071–1086 (2009)
26. Markovich-Golan, S., Gannot, S.: Performance analysis of the covariance subtraction method for relative transfer function estimation and comparison to the covariance whitening method. In: Proceedings of the IEEE International Conference on Acoustics, Speech and Signal Processing (ICASSP), pp. 544–548 (2015). doi:10.1109/ICASSP.2015.7178028
27. Markovich-Golan, S., Gannot, S., Cohen, I.: A sparse blocking matrix for multiple constraints GSC beamformer. In: Proceedings of the IEEE International Conference on Acoustics, Speech and Signal Processing (ICASSP), pp. 197–200 (2012)
28. Meyer, J., Elko, G.: A highly scalable spherical microphone array based on an orthonormal decomposition of the soundfield. In: Proceedings of the IEEE International Conference on Acoustics, Speech and Signal Processing (ICASSP), vol. 2, pp. 1781–1784 (2002)
29. Nordholm, S., Claesson, I., Eriksson, P.: The broadband Wiener solution for Griffiths-Jim beamformers. IEEE Trans. Signal Process. **40**(2), 474–478 (1992)
30. Peled, Y., Rafaely, B.: Linearly constrained minimum variance method for spherical microphone arrays in a coherent environment. In: Proceedings of the Hands-Free Speech Communication and Microphone Arrays (HSCMA), pp. 86–91 (2011). doi:10.1109/HSCMA.2011.5942416
31. Rafaely, B.: Plane-wave decomposition of the pressure on a sphere by spherical convolution. J. Acoust. Soc. Am. **116**(4), 2149–2157 (2004)
32. Shalvi, O., Weinstein, E.: System identification using nonstationary signals. IEEE Trans. Signal Process. **44**(5), 2055–2063 (1996)
33. Souden, M., Benesty, J., Affes, S.: On optimal frequency-domain multichannel linear filtering for noise reduction. IEEE Trans. Audio, Speech, Lang. Process. **18**(2), 260–276 (2010). http://dx.doi.org/10.1109/TASL.2009.2025790
34. Teutsch, H.: Wavefield decomposition using microphone arrays and its application to acoustic scene analysis. Ph.D. thesis, Friedrich-Alexander Universität Erlangen-Nürnberg (2005)
35. van Trees, H.L.: Detection, Estimation, and Modulation Theory Optimum Array Processing, vol. IV. Wiley, New York (2002)
36. van Trees, H.L.: Optimum Array Processing. Detection, Estimation and Modulation Theory. Wiley, New York (2002)

37. van Veen, B.D., Buckley, K.M.: Beamforming: a versatile approach to spatial filtering. IEEE Acoust. Speech Signal Mag. **5**(2), 4–24 (1988)
38. Wiener, N.: The Extrapolation, Interpolation and Smoothing of Stationary Time Series. Wiley Inc., New York (1949)
39. Williams, E.G.: Fourier Acoustics: Sound Radiation and Nearfield Acoustical Holography, 1st edn. Academic Press, London (1999)
40. Yan, S., Sun, H., Svensson, U.P., Ma, X., Hovem, J.M.: Optimal modal beamforming for spherical microphone arrays. IEEE Trans. Audio, Speech, Lang. Process. **19**(2), 361–371 (2011). doi:10.1109/TASL.2010.2047815

Chapter 8
Parametric Array Processing

The general principle of parametric array processing is to employ an efficient parametric representation of the sound field including typically one or a few reference signals, and a small number of associated parameters. The advantage of such an approach is that the number of parameters is significantly lower than in classical array processing (see Chap. 7). A block diagram of a parametric array processing approach is shown in Fig. 8.1.

Examples of parametric representations of the sound field include Directional Audio Coding (DirAC) [11], High Angular Resolution Planewave Expansion (HARPEX) [1] and computational auditory scene analysis (CASA) [4]. These representations can be used for spatial audio recording, coding and reproduction; source separation, noise reduction and dereverberation; and acoustic scene analysis and source localization. In this chapter, we will focus on parametric approaches to signal enhancement using the DirAC representation.

The DirAC representation is based on two features that are relevant to the perception of spatial sound: the direction of arrival (DOA) and the diffuseness. Providing these features are accurately reproduced, this representation ensures that the interaural time differences (ITDs), interaural level differences (ILDs), and the interaural coherence are correctly perceived [16]. The advantage of integrating DirAC with a signal enhancement process is that any interference sources can be reproduced at their original position [9] relative to the desired source, in addition to being attenuated, thereby maintaining the naturalness of the listening experience but with increased speech quality and intelligibility.

In this chapter, we first introduce a parametric model of the sound field. We then review the parameters that describe this sound field and how they can be estimated, and present filters that can be used to separate the two components of the sound field. Finally, we explore two applications of parametric array processing, namely, directional filtering and dereverberation.

D. Jarrett et al., *Theory and Applications of Spherical Microphone Array Processing*, Springer Topics in Signal Processing 9,
DOI 10.1007/978-3-319-42211-4_8

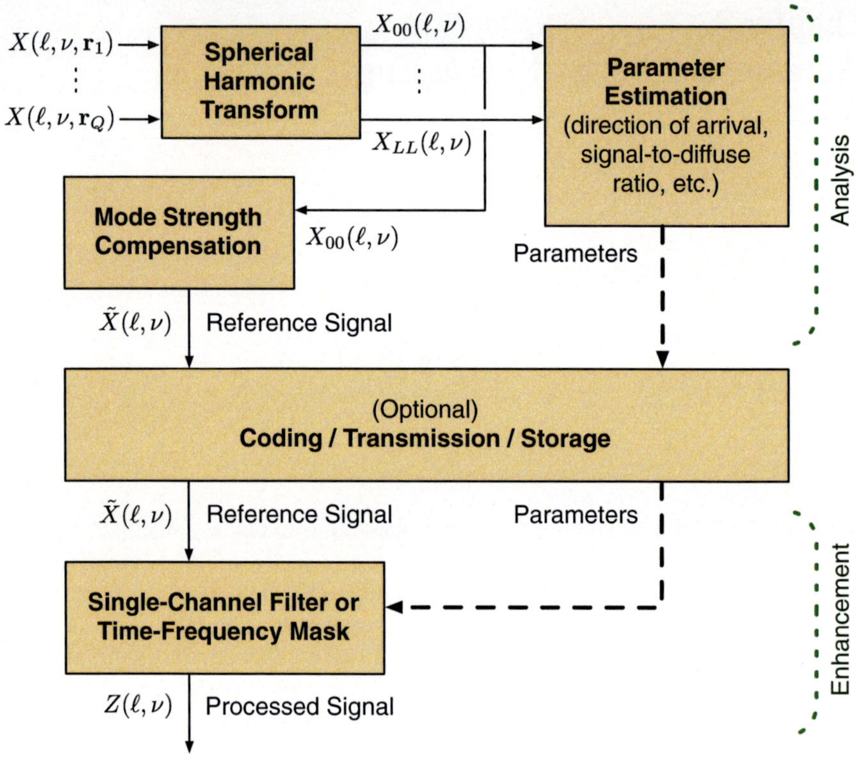

Fig. 8.1 Block diagram of a parametric array processing approach. In the analysis stage, a reference signal is computed and a number of parameters are estimated. The reference signal and estimated parameters are transmitted or stored. In the enhancement stage, a single-channel filter or time-frequency mask is applied to the reference signal, optionally based on the estimated parameters, to yield a processed output signal

8.1 Signal Model

In the short-time Fourier transform (STFT) domain, the sound pressure S at a position $\mathbf{r}$ can be decomposed into a direct sound component S_{dir} and a diffuse sound component S_{diff}, such that

$$S(\ell, \nu, \mathbf{r}) = S_{\text{dir}}(\ell, \nu, \mathbf{r}) + S_{\text{diff}}(\ell, \nu, \mathbf{r}), \tag{8.1}$$

where ℓ denotes the discrete time index and ν denotes the discrete frequency index. The sound pressure signal X measured by Q microphones at positions $\mathbf{r}_q, q \in \{1, \ldots, Q\}$ is then given by

$$X(\ell, \nu, \mathbf{r}_q) = S(\ell, \nu, \mathbf{r}_q) + V(\ell, \nu, \mathbf{r}_q) \tag{8.2a}$$
$$= S_{\text{dir}}(\ell, \nu, \mathbf{r}_q) + S_{\text{diff}}(\ell, \nu, \mathbf{r}_q) + V(\ell, \nu, \mathbf{r}_q), \tag{8.2b}$$

where V denotes a sensor noise signal.

We assume that the directional signal S_{dir} is sparse in the time-frequency domain [12], such that in each time-frequency bin the directional signal is due to a single plane wave. The diffuse signal is due to a theoretically infinite number of independent plane waves with random phases, equal amplitudes and uniformly distributed DOAs [10]. We also assume that all three signals are mutually uncorrelated, that is,

$$\mathrm{E}\left\{S_{\text{dir}}(\ell, \nu, \mathbf{r}_q) S^*_{\text{diff}}(\ell, \nu, \mathbf{r}_q)\right\} = 0 \tag{8.3}$$
$$\mathrm{E}\left\{S_{\text{dir}}(\ell, \nu, \mathbf{r}_q) V^*(\ell, \nu, \mathbf{r}_q)\right\} = 0, \tag{8.4}$$

where $\mathrm{E}\{\cdot\}$ denotes mathematical expectation, which can be computed using temporal averaging.

In order to obtain the reference signal as indicated in Fig. 8.1, we must transform the spatial domain signals to the spherical harmonic domain (SHD). In this chapter, we assume error-free spatial sampling, and refer the reader to Chap. 3 for information on spatial sampling and aliasing. By applying the complex spherical harmonic transform (SHT) to the signal model in (8.2), we obtain the SHD signal model

$$X_{lm}(\ell, \nu) = S_{lm}(\ell, \nu) + V_{lm}(\ell, \nu) \tag{8.5a}$$
$$= S^{\text{dir}}_{lm}(\ell, \nu) + S^{\text{diff}}_{lm}(\ell, \nu) + V_{lm}(\ell, \nu), \tag{8.5b}$$

where $X_{lm}(\ell, \nu)$, $S_{lm}(\ell, \nu)$, $S^{\text{dir}}_{lm}(\ell, \nu)$, $S^{\text{diff}}_{lm}(\ell, \nu)$ and $V_{lm}(\ell, \nu)$ are respectively the spherical harmonic transforms of the signals $X(\ell, \nu, \mathbf{r}_q)$, $S(\ell, \nu, \mathbf{r}_q)$, $S_{\text{dir}}(\ell, \nu, \mathbf{r}_q)$, $S_{\text{diff}}(\ell, \nu, \mathbf{r}_q)$ and $V(\ell, \nu, \mathbf{r}_q)$, as defined in (3.6), and are referred to as *eigenbeams* to reflect the fact that the spherical harmonics are eigensolutions of the wave equation in spherical coordinates [14]. The order and degree of the spherical harmonics are respectively denoted as l and m.

We choose as a reference the signal that would be measured by an omnidirectional microphone $\mathcal{M}_{\text{ref}}$ placed at the centre of the spherical array, if the array were not present. As shown in the Appendix of Chap. 5, the sound pressure $\widetilde{X}(\ell, \nu)$ at this microphone can be obtained from the zero-order eigenbeam $X_{00}(\ell, \nu)$ as

$$\widetilde{X}(\ell, \nu) = \frac{X_{00}(\ell, \nu)}{\sqrt{4\pi} B_0(\nu)} \tag{8.6a}$$
$$= \widetilde{S}(\ell, \nu) + \widetilde{V}(\ell, \nu) \tag{8.6b}$$
$$= \widetilde{S}_{\text{dir}}(\ell, \nu) + \widetilde{S}_{\text{diff}}(\ell, \nu) + \widetilde{V}(\ell, \nu), \tag{8.6c}$$

where the frequency-dependent mode strength $B_l(\nu)$ for spherical harmonic order l, given by evaluating the wavenumber-dependent mode strength $b_l(k)$ at discrete values of the wavenumber k, captures the dependence of the lth order eigenbeams on the array properties, and is discussed in Sect. 3.4.2. By dividing the eigenbeam $X_{00}(\ell, \nu)$ by the mode strength, we remove this dependence, such that the reference signal is independent of the array properties. As noted in Sect. 7.2.2, assuming the array's Q microphones are uniformly distributed on the sphere, the power of the sensor noise V is $Q\,|B_0(\nu)|^2$ times smaller at $\mathcal{M}_{\text{ref}}$ than at the individual microphones on the surface of the sphere.

The directional signal S_{lm}^{dir} due to a plane wave incident from a direction Ω_{dir} is given by

$$S_{lm}^{\text{dir}}(\ell, \nu) = \sqrt{P_{\text{dir}}(\ell, \nu)}\varphi_{\text{dir}}(\ell, \nu) 4\pi B_l(\nu) Y_{lm}^{*}\left[\Omega_{\text{dir}}(\ell, \nu)\right], \tag{8.7}$$

where $\varphi_{\text{dir}}(\ell, \nu)$ is the phase factor of the plane wave, $P_{\text{dir}}(\ell, \nu)$ is the power of the plane wave, and Y_{lm} is the complex spherical harmonic,[1] as defined in (2.14). The diffuse signal S_{lm}^{dir} can be expressed as

$$S_{lm}^{\text{diff}}(\ell, \nu) = \sqrt{\frac{P_{\text{diff}}(\ell, \nu)}{4\pi}} \int_{\Omega \in \mathcal{S}^2} \varphi_{\text{diff}}(\ell, \nu, \Omega) 4\pi B_l(\nu) Y_{lm}^{*}(\Omega) \mathrm{d}\Omega, \tag{8.8}$$

where $\varphi_{\text{diff}}(\ell, \nu, \Omega)$ denotes the phase factor of the plane wave incident from direction Ω and the notation $\int_{\Omega \in \mathcal{S}^2} \mathrm{d}\Omega$ is used to denote compactly the solid angle $\int_{\phi=0}^{2\pi} \int_{\theta=0}^{\pi} \sin\theta \mathrm{d}\theta \mathrm{d}\phi$.

As in Sect. 5.2.1, using the relationship (5.74) between the zero-order eigenbeam $X_{00}(\ell, \nu)$ and the reference signal $\widetilde{X}(\ell, \nu)$, as well as the expressions for the directional and diffuse signals in (8.7) and (8.8), it can be verified that the powers of these signals at $\mathcal{M}_{\text{ref}}$ are respectively given by P_{dir} and P_{diff}.

8.2 Parameter Estimation

In the parametric model, the sound field is described by two parameters for each time-frequency bin: the DOA $\Omega_{\text{dir}}(\ell, \nu)$ of the plane wave that generates the directional signal, and the diffuseness $\Psi(\ell, \nu)$, which determines the strength of the directional signal with respect to the diffuse signal.

The diffuseness is defined as [5]

$$\Psi(\ell, \nu) = \frac{1}{1 + \Gamma(\ell, \nu)}, \tag{8.9}$$

[1] If the real SHT is applied instead of the complex SHT, the complex spherical harmonics Y_{lm} used throughout this chapter should be replaced with the real spherical harmonics R_{lm}, as defined in Sect. 3.3.

where $\Gamma(\ell, \nu)$ denotes the signal-to-diffuse ratio (SDR) at $\mathcal{M}_{\text{ref}}$, given by

$$\Gamma(\ell, \nu) = \frac{|\widetilde{S}_{\text{dir}}(\ell, \nu)|^2}{\mathrm{E}\left\{|\widetilde{S}_{\text{diff}}(\ell, \nu)|^2\right\}} \tag{8.10a}$$

$$= \frac{|S_{00}^{\text{dir}}(\ell, \nu)|^2}{\mathrm{E}\left\{|S_{00}^{\text{diff}}(\ell, \nu)|^2\right\}} \tag{8.10b}$$

$$= \frac{P_{\text{dir}}(\ell, \nu)}{P_{\text{diff}}(\ell, \nu)}. \tag{8.10c}$$

The diffuseness takes values between 0 and 1. In a purely directional field, a diffuseness of 0 is obtained; in a purely directional field, a diffuseness of 1 is obtained; and when the directional and diffuse signals have equal power, a diffuseness of 0.5 is obtained.

Time- and frequency-dependent DOA and SDR/diffuseness estimates can be obtained using the methods presented in Chap. 5. In order for the reproduction of the sound field to be accurate, and to avoid distortion of the signals when enhancement is performed, it is crucial that the parameter estimates have sufficiently high temporal and spectral resolution, as well as sufficiently low variance.

8.3 Sound Pressure Estimation

In order to perform signal enhancement, we would like to estimate the directional and diffuse components $\widetilde{S}_{\text{dir}}(\ell, \nu)$ and $\widetilde{S}_{\text{diff}}(\ell, \nu)$ of the reference signal $\widetilde{X}(\ell, \nu)$. This can be done by applying a square-root Wiener filter to $\widetilde{X}(\ell, \nu)$, such that

$$\hat{S}_{\text{dir}}(\ell, \nu) = W_{\text{dir}}(\ell, \nu)\widetilde{X}(\ell, \nu) \tag{8.11}$$

$$\hat{S}_{\text{diff}}(\ell, \nu) = W_{\text{diff}}(\ell, \nu)\widetilde{X}(\ell, \nu), \tag{8.12}$$

where the directional filter weights are given by

$$W_{\text{dir}}(\ell, \nu) = \sqrt{\frac{P_{\text{dir}}(\ell, \nu)}{P_{\text{dir}}(\ell, \nu) + P_{\text{diff}}(\ell, \nu) + \mathrm{E}\left\{|\widetilde{V}(\ell, \nu)|^2\right\}}} \tag{8.13a}$$

$$= \sqrt{\frac{\Gamma(\ell, \nu)}{\Gamma(\ell, \nu) + 1 + P_{\text{diff}}^{-1}(\ell, \nu)\,\mathrm{E}\left\{|\widetilde{V}(\ell, \nu)|^2\right\}}} \tag{8.13b}$$

and the diffuse filter weights are given by

$$W_{\text{diff}}(\ell,\nu) = \sqrt{\frac{P_{\text{diff}}(\ell,\nu)}{P_{\text{dir}}(\ell,\nu) + P_{\text{diff}}(\ell,\nu) + \mathrm{E}\left\{|\widetilde{V}(\ell,\nu)|^2\right\}}} \tag{8.14a}$$

$$= \sqrt{\frac{1}{\Gamma(\ell,\nu) + 1 + P_{\text{diff}}^{-1}(\ell,\nu)\,\mathrm{E}\left\{|\widetilde{V}(\ell,\nu)|^2\right\}}}. \tag{8.14b}$$

Because the power of the spatially incoherent sensor noise is reduced when combining the Q microphone signals, we can assume that the power of the sensor noise $\widetilde{V}(\ell,\nu)$ is negligible, and therefore $\mathrm{E}\left\{|\widetilde{V}(\ell,\nu)|^2\right\} = 0$. In this case, the filter weights can be simplified to

$$W_{\text{dir}}(\ell,\nu) = \sqrt{\frac{\Gamma(\ell,\nu)}{\Gamma(\ell,\nu)+1}} \tag{8.15a}$$

$$= \sqrt{1 - \Psi(\ell,\nu)} \tag{8.15b}$$

and

$$W_{\text{diff}}(\ell,\nu) = \sqrt{\frac{1}{\Gamma(\ell,\nu)+1}} \tag{8.16a}$$

$$= \sqrt{\Psi(\ell,\nu)} \tag{8.16b}$$

$$= \sqrt{1 - W_{\text{dir}}^2(\ell,\nu)}. \tag{8.16c}$$

If the sensor noise power is not sufficiently low to be disregarded, the filter weights can be computed using an estimate of the diffuse-to-noise ratio, obtained using the method in [15], for example.

The advantage of using a square-root Wiener filter in this context is that the power of the directional and diffuse signals is preserved, that is, $\mathrm{E}\left\{|\hat{S}_{\text{dir}}(\ell,\nu)|^2\right\} = P_{\text{dir}}(\ell,\nu)$ and $\mathrm{E}\left\{|\hat{S}_{\text{diff}}(\ell,\nu)|^2\right\} = P_{\text{diff}}(\ell,\nu)$. In practice, a lower bound is sometimes applied to W_{dir} in order to avoid introducing audible artefacts such as musical noise [2, 18]. In addition, if the diffuse filter weights are computed using (8.16c), $\mathrm{E}\left\{|\hat{S}_{\text{dir}}(\ell,\nu)|^2\right\} + \mathrm{E}\left\{|\hat{S}_{\text{diff}}(\ell,\nu)|^2\right\} = \mathrm{E}\left\{|\widetilde{X}(\ell,\nu)|^2\right\}$, even if a lower bound is applied to W_{dir}.

8.4 Applications

In this section, we consider two applications of parametric array processing to signal enhancement: directional filtering (Sect. 8.4.1) and dereverberation (Sect. 8.4.2). The general principle in both of these applications is to apply a single-channel filter or time-frequency mask to the reference signal $\widetilde{X}(\ell, \nu)$ or the estimated pressure signals $\hat{S}_{\text{dir}}(\ell, \nu)$ and $\hat{S}_{\text{diff}}(\ell, \nu)$. As well as enhancing the signal, this can unfortunately also introduce speech distortion or musical noise, especially with filters that vary quickly across time and frequency. However, this problem can be mitigated by establishing a lower bound on the filter weights (as in Sect. 8.3), or by smoothing the weights across time and frequency [3, 7].

8.4.1 Directional Filtering

As proposed by Kallinger et al. in [8], a directional filter can be implemented by modifying the reference signal $\widetilde{X}(\ell, \nu)$, the diffuseness $\Psi(\ell, \nu)$ and the DOA $\Omega_{\text{dir}}(\ell, \nu)$. In this section, we apply two filters $W_{\text{dir}}^{\text{filt}}$ and $W_{\text{diff}}^{\text{filt}}$ directly to the estimated direct and diffuse sound pressures, such that

$$Z_{\text{dir}}(\ell, \nu) = W_{\text{dir}}^{\text{filt}}\left[\Omega(\ell, \nu)\right] \hat{S}_{\text{dir}}(\ell, \nu) \tag{8.17}$$

$$Z_{\text{diff}}(\ell, \nu) = W_{\text{diff}}^{\text{filt}}\, \hat{S}_{\text{diff}}(\ell, \nu). \tag{8.18}$$

The filtered reference signal is then given by summing the filtered directional and diffuse signals:

$$Z(\ell, \nu) = Z_{\text{dir}}(\ell, \nu) + Z_{\text{diff}}(\ell, \nu). \tag{8.19}$$

We would like the filtered reference signal to correspond to the signal captured by a directional microphone with a directional response $D\left[\Omega\right]$. We additionally want a directional response of unity in the microphone's steering direction Ω_{u}. Ideally, we would be able to use a Dirac delta function in the steering direction. However, in practice this is not possible because the DOA estimates are not error-free and the directional sources are not point sources [8]. In practice, a beam width in the region of 60° can be achieved without introducing significant audible artefacts [8].

We can choose, for example, a first-order microphone steered in a direction $\Omega_{\text{u}} = (\theta_{\text{u}}, \phi_{\text{u}})$, whose directional response is given by [6]

$$\begin{aligned} D\left[\Omega(\ell, \nu)\right] = \alpha + (1-\alpha)\Big\{ &\sin\left[\theta(\ell, \nu)\right] \sin\theta_{\text{u}} \cos\left[\phi(\ell, \nu) - \phi_{\text{u}}\right] \\ &+ \cos\left[\theta(\ell, \nu)\right] \cos\theta_{\text{u}} \Big\}, \end{aligned} \tag{8.20}$$

where the term in curly brackets is the cosine of the angle between the DOA $\Omega = (\theta, \phi)$ and steering direction Ω_{u}, and α is a shape parameter for the first-order microphone. In Table 8.1, we list a number of commonly used directivity patterns and the corresponding shape parameters.

The power of an ideal diffuse signal (with unit power at $\mathcal{M}_{\text{ref}}$) at the output of such a microphone is given by [6, 17]

Table 8.1 Commonly used first-order directivity patterns and corresponding shape parameter values

Directivity pattern	Shape parameter α
Omnidirectional	1
Subcardioid	0.75
Cardioid	0.5
Hypercardioid	0.25
Bidirectional	0

$$P_{D_{\mathrm{diff}}} = \frac{1}{4\pi} \int_{\Omega \in \mathcal{S}^2} D^2 \left[\Omega(\ell, \nu)\right] \mathrm{d}\Omega \tag{8.21a}$$

$$= \frac{4}{3}\alpha^2 - \frac{2}{3}\alpha + \frac{1}{3}. \tag{8.21b}$$

The directional and diffuse filter weights are then given by

$$W_{\mathrm{dir}}^{\mathrm{filt}} \left[\Omega(\ell, \nu)\right] = D\left[\Omega(\ell, \nu)\right] \tag{8.22}$$

$$W_{\mathrm{diff}}^{\mathrm{filt}} = \sqrt{P_{D_{\mathrm{diff}}}}. \tag{8.23}$$

This directional filtering technique can be likened to beamforming, and indeed the objective is the same. However, this technique involves a single-channel filter, while in beamforming we apply a filter to the pressure signals recorded at multiple microphones, or to multiple eigenbeams.

8.4.2 *Dereverberation*

In [9], Kallinger et al. also proposed a method for dereverberation using a parametric approach. The desired signal that contains less reverberation than the reference signal $\widetilde{X}(\ell, \nu)$ is given by

$$\widetilde{X}_{\mathrm{dereverb}}(\ell, \nu) = S_{\mathrm{dir}}(\ell, \nu) + \beta(\ell, \nu) S_{\mathrm{diff}}(\ell, \nu), \tag{8.24}$$

where $0 \leq \beta(\ell, \nu) < 1$ is a reverberation reduction factor.

A single-channel filter $W(\ell, \nu)$ can be applied to the reference signal $\widetilde{X}(\ell, \nu)$ to estimate the desired signal $\widetilde{X}_{\mathrm{dereverb}}(\ell, \nu)$:

$$Z(\ell, \nu) = W(\ell, \nu)\widetilde{X}(\ell, \nu). \tag{8.25}$$

The filter weights $W_{\mathrm{MMSE}}(\ell, \nu)$ that minimize the mean square error between the filter output signal $Z(\ell, \nu)$ and the desired signal $\widetilde{X}_{\mathrm{dereverb}}(\ell, \nu)$ are given by

$$
\begin{aligned}
W_{\text{MMSE}}(\ell,\nu) &= \underset{W(\ell,\nu)}{\arg\min}\, \mathrm{E}\left\{\left|\widetilde{X}_{\text{dereverb}}(\ell,\nu) - W(\ell,\nu)\widetilde{X}(\ell,\nu)\right|^2\right\} && \text{(8.26a)}\\
&= 1-(1-\beta)\Psi(\ell,\nu) && \text{(8.26b)}\\
&= \frac{\Gamma(\ell,\nu)+\beta(\ell,\nu)}{\Gamma(\ell,\nu)+1}. && \text{(8.26c)}
\end{aligned}
$$

This filter is attractive due to its simplicity, since the filter weights only depend on the diffuseness and the desired reverberation reduction factor and do not depend on the DOA. As previously mentioned, the filter weights must normally be smoothed over time and frequency to avoid audible artefacts; the amount of smoothing that is necessary will depend on how much smoothing has been applied to the diffuseness estimates.

It should be noted that the filter described in this section can be used to suppress any diffuse sound, whether it be reverberation, or isotropic noise such as car noise or babble noise.

8.5 Chapter Summary

Parametric array processing relies on a simple yet powerful parametric model of the sound field, which in this chapter was described using a single reference pressure signal along with two parameters, the DOA and the diffuseness. These parameters must be estimated accurately, and with high time and frequency resolution. We presented two illustrative applications of this array processing approach: directional filtering and dereverberation. These applications highlight a significant advantage of parametric array processing techniques: they typically have low computational complexity, especially if low-complexity parameter estimation methods are chosen (see Chap. 5).

Ongoing research challenges include formulating more sophisticated parametric models to improve performance, and finding new ways to avoid audible artefacts despite using filters whose weights vary quickly with time and frequency. Other potential applications of parametric array processing include acoustic zoom [13, 19] and source extraction using multiple microphone arrays.

References

1. Berge, S., Barrett, N.: High angular resolution planewave expansion. In: Proceedings of the 2nd International Symposium on Ambisonics and Spherical Acoustics (2010)
2. Berouti, M., Schwartz, R., Makhoul, J.: Enhancement of speech corrupted by acoustic noise. In: Proceedings of the IEEE International Conference on Acoustics, Speech and Signal Processing (ICASSP), vol. 4, pp. 208–211 (1979)
3. Breithaupt, C., Gerkmann, T., Martin, R.: Cepstral smoothing of spectral filter gains for speech enhancement without musical noise. IEEE Signal Process. Lett. **14**(12), 1036–1039 (2007)

4. Brown, G.J., Cooke, M.: Computational auditory scene analysis. Comput. Speech Lang. **8**, 297–336 (1994)
5. Del Galdo, G., Taseska, M., Thiergart, O., Ahonen, J., Pulkki, V.: The diffuse sound field in energetic analysis. J. Acoust. Soc. Am. **131**(3), 2141–2151 (2012)
6. Elko, G.W.: Spatial coherence functions for differential microphones in isotropic noise fields. In: Brandstein, M., Ward, D. (eds.) Microphone Arrays: Signal Processing Techniques and Applications, chap. 4, pp. 61–85. Springer, Heidelberg (2001)
7. Gustafsson, S., Nordholm, S., Claesson, I.: Spectral subtraction using reduced delay convolution and adaptive averaging. IEEE Trans. Speech Audio Process. **9**(8), 799–807 (2001)
8. Kallinger, M., Ochsenfeld, H., Del Galdo, G., Kuech, F., Mahne, D., Schultz-Amling, R., Thiergart, O.: A spatial filtering approach for directional audio coding. In: Proceedings of the Audio Engineering Society Convention. Munich, Germany (2009)
9. Kallinger, M., Del Galdo, G., Kuech, F., Thiergart, O.: Dereverberation in the spatial audio coding domain. In: Proceedings of the Audio Engineering Society Convention. London, UK (2011)
10. Kuttruff, H.: Room Acoustics, 4th edn. Taylor & Francis, London (2000)
11. Pulkki, V.: Spatial sound reproduction with directional audio coding. J. Audio Eng. Soc. **55**(6), 503–516 (2007)
12. Rickard, S., Yilmaz, Z.: On the approximate W-disjoint orthogonality of speech. In: Proceedings of the IEEE International Conference on Acoustics, Speech and Signal Processing (ICASSP), vol. 1, pp. 529–532 (2002)
13. Schultz-Amling, R., Kuech, F., Thiergart, O., Kallinger, M.: Acoustical zooming based on a parametric sound field representation. In: Proceedings of the Audio Engineering Society Convention (2010)
14. Teutsch, H.: Wavefield decomposition using microphone arrays and its application to acoustic scene analysis. Ph.D. thesis, Friedrich-Alexander Universität Erlangen-Nürnberg (2005)
15. Thiergart, O., Habets, E.A.P.: An informed LCMV filter based on multiple instantaneous direction-of-arrival estimates. In: Proceedings of the IEEE International Conference on Acoustics, Speech and Signal Processing (ICASSP), pp. 659–663 (2013)
16. Thiergart, O., Kallinger, M., Del Galdo, G., Kuech, F.: Parametric spatial sound processing using linear microphone arrays. In: Heuberger, A., Elst, G., Hanke, R. (eds.) Microelectronic Systems, pp. 313–321. Springer, Heidelberg (2011)
17. Thiergart, O., Del Galdo, G., Habets, E.A.P.: On the spatial coherence in mixed sound fields and its application to signal-to-diffuse ratio estimation. J. Acoust. Soc. Am. **132**(4), 2337–2346 (2012)
18. Thiergart, O., Del Galdo, G., Taseska, M., Habets, E.: Geometry-based spatial sound acquisition using distributed microphone arrays. IEEE Trans. Audio, Speech, Lang. Process. **21**(12), 2583–2594 (2013)
19. Thiergart, O., Kowalczyk, K., Habets, E.: An acoustical zoom based on informed spatial filtering. In: Proceedings of the International Workshop Acoustic Signal Enhancement (IWAENC), pp. 109–113. IEEE, Juan-les-Pins, France (2014). doi:10.1109/IWAENC.2014.6953348

Chapter 9
Informed Array Processing

Classical beamformers allow us to control the tradeoff between noise reduction and speech distortion, but are not very robust to estimation errors and source position changes and have a slow response time. In contrast, parametric spatial filtering techniques have a fast response time and are relatively robust, but do not allow us to control this tradeoff, can suffer from audible artefacts when the parametric model is violated, and have relatively poor interference reduction.

Informed array processing aims to bridge the gap between these two approaches. The conceptual aim of informed array processing is to incorporate relevant information about the problem to be solved into the design of spatial filters and into the estimation of the second-order statistics that are required to implement these filters.

The information that can be used to inform the design of the filter weights and the statistical estimation includes time- and frequency-dependent

- signal-to-diffuse ratio (SDR) estimates (obtained using the algorithms in Sect. 5.2, for example);
- direction of arrival (DOA) estimates (obtained using the algorithms in Sect. 5.1, for example);
- interaural time difference (ITD) or interaural level difference (ILD) estimates; and
- position estimates (this typically requires multiple arrays).

The informed array processing approach is illustrated in the form of a block diagram in Fig. 9.1, where an informed spatial filter is applied to the spherical harmonic domain (SHD) pressure signals, the eigenbeams, to obtain an enhanced output signal. The eigenbeams are also used to estimate acoustic parameters, which are then used to estimate second-order statistics and (optionally) compute the weights of the informed spatial filter.

D. Jarrett et al., *Theory and Applications of Spherical Microphone Array Processing*, Springer Topics in Signal Processing 9,
DOI 10.1007/978-3-319-42211-4_9

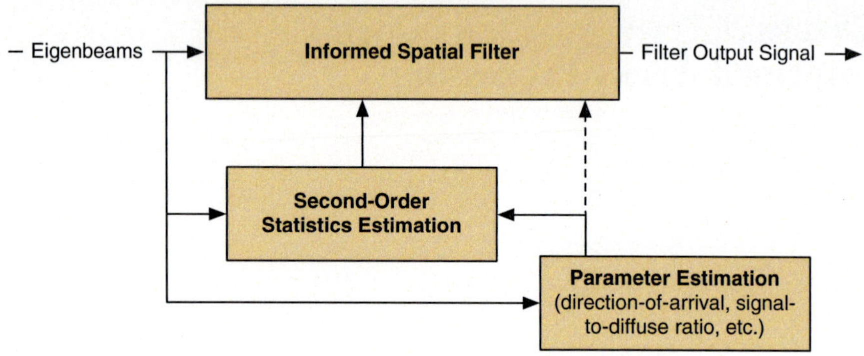

Fig. 9.1 Block diagram of an informed spatial filtering approach

This approach can be applied to problems such as noise reduction, dereverberation or source extraction. In this chapter, we look at two application scenarios: coherent and incoherent noise reduction (Sect. 9.1) using instantaneous DOA estimates, and joint dereverberation and incoherent noise reduction using instantaneous SDR estimates (Sect. 9.2).

9.1 Noise Reduction Using Narrowband DOA Estimates

The implementation of the SHD signal-dependent beamformers presented in Chap. 7 requires the estimation of the second-order statistics of the desired and noise signals, most importantly the power spectral density (PSD) matrix of the noise. Unfortunately, in practice this is not a straightforward problem since the desired and noise signals cannot be observed directly, and their statistics must instead be estimated from the noisy signals.

In the spatial domain, the noise PSD matrix has previously been estimated based on the speech presence probability (SPP) [13, 15, 40]. The noise PSD estimate is then only updated in time-frequency bins where speech is likely to be absent, similarly to single-channel approaches [4, 8]. Souden et al. proposed a Gaussian model–based multichannel SPP estimator [39] that can detect spatially coherent sources regardless of their DOA.

In order to perform coherent noise reduction, it can be desirable to distinguish between desired coherent sources, located within a given region of interest, and undesired coherent sources, which are considered as noise sources. This distinction cannot be made using only the SPP and requires us to draw on additional spatial information. In [24], the authors proposed to estimate the PSD matrices of the desired and noise signals using a desired speech presence probability (DSPP), given by the product of Souden et al.'s multichannel SPP and a DOA-based probability. The latter probability is based on DOA estimates for each time-frequency bin: using a priori information on the variance of the DOA estimates, the probability that the active coherent source lies within the region of interest can be determined. It is assumed that the bounds of the region of interest are known a priori; they can be chosen manually, or estimated using visual information such as face tracking [45].

The second-order statistics thereby estimated can then be used to compute the weights of a tradeoff beamformer, as introduced in Sect. 7.3.4, which seeks to balance noise reduction against speech distortion based on a tradeoff parameter that can optionally be DSPP-dependent. The tradeoff beamformer and second-order statistics estimation method form a complete informed noise reduction algorithm.

9.1.1 Signal Models

Spatial Domain Signal Model

We consider a short-time Fourier transform (STFT) domain signal model in which a Q-microphone spherical array of radius r captures the sound pressure $P(\ell, \nu, \mathbf{r}_q)$ at positions $\mathbf{r}_q = (r, \Omega_q) = (r, \theta_q, \phi_q), q \in \{1, \ldots, Q\}$ (in spherical coordinates, where θ_q denotes the inclination and ϕ_q denotes the azimuth), where ℓ denotes the discrete time index[1] and ν denotes the discrete frequency index.

The sound field is composed of a mixture of desired speech, originating from a desired source $\mathcal{S}$; spatially coherent noise arising from, for example, interfering speech; and background noise. The background noise may consist of a mixture of spatially incoherent noise, used to model sensor noise, and partially coherent noise, used to model spherically or cylindrically isotropic noise. The reader is referred to Sect. 1.2 for a discussion on the spatial characteristics of sound fields.

The signal model that corresponds to this scenario can be expressed in the STFT domain as

$$\begin{aligned} P(\nu, \mathbf{r}_q) &= H(\nu, \mathbf{r}_q) S(\nu) + V_{\mathrm{c}}(\nu, \mathbf{r}_q) + V_{\mathrm{nc}}(\nu, \mathbf{r}_q) \\ &= X(\nu, \mathbf{r}_q) + V_{\mathrm{c}}(\nu, \mathbf{r}_q) + V_{\mathrm{nc}}(\nu, \mathbf{r}_q), \end{aligned} \tag{9.1}$$

where S is the source signal produced by the desired source $\mathcal{S}$, X is the convolved source signal originating from the source $\mathcal{S}$, V_{c} is the coherent noise signal, V_{nc} is the background noise signal, and $H(\nu, \mathbf{r}_q)$ is the acoustic transfer function (ATF) between the source $\mathcal{S}$ and the microphone at position $\mathbf{r}_q$. The desired source $\mathcal{S}$ is located inside a region of interest $\mathcal{R}$, whereas the coherent noise source(s) are located outside the region $\mathcal{R}$.

We assume the reverberant speech signals $X(\nu, \mathbf{r}_q)$ and the noise signals $V_{\mathrm{c}}(\nu, \mathbf{r}_q)$ and $V_{\mathrm{nc}}(\nu, \mathbf{r}_q)$ are mutually uncorrelated. As the reverberant speech signals $X(\nu, \mathbf{r}_q)$ originate from a single source, they are, by definition, coherent at all microphones in the array.

[1] For brevity, the dependency of all quantities on ℓ is omitted throughout Sects. 9.1.1 and 9.1.2.

Spherical Harmonic Domain Signal Model

When using spherical microphone arrays, it is convenient to work in the SHD [28, 36], instead of the spatial domain. In this chapter, we assume error-free spatial sampling, and refer the reader to Chap. 3 for information on spatial sampling and aliasing. By applying the complex spherical harmonic transform (SHT) to the signal model in (9.1), we obtain the SHD signal model

$$\begin{aligned} P_{lm}(\nu) &= H_{lm}(\nu)S(\nu) + V_{lm,\text{c}}(\nu) + V_{lm,\text{nc}}(\nu) \\ &= X_{lm}(\nu) + V_{lm,\text{c}}(\nu) + V_{lm,\text{nc}}(\nu), \end{aligned} \tag{9.2}$$

where $P_{lm}(\nu)$, $H_{lm}(\nu)$, $X_{lm}(\nu)$, $V_{lm,\text{c}}(\nu)$ and $V_{lm,\text{nc}}(\nu)$ are respectively the spherical harmonic transforms of the signals $P(\nu, \mathbf{r}_q)$, $H(\nu, \mathbf{r}_q)$, $X(\nu, \mathbf{r}_q)$, $V_\text{c}(\nu, \mathbf{r}_q)$ and $V_\text{nc}(\nu, \mathbf{r}_q)$, as defined in (3.6), and are referred to as *eigenbeams* to reflect the fact that the spherical harmonics are eigensolutions of the wave equation in spherical coordinates [44]. The order and degree of the spherical harmonics are respectively denoted as l and m.

Mode Strength Compensation

The eigenbeams P_{lm}, H_{lm}, X_{lm}, $V_{lm,\text{c}}$ and $V_{lm,\text{nc}}$ are a function of the mode strength $B_l(\nu)$, which is in turn a function of the array properties (radius, microphone type, configuration). The mode strength $B_l(\nu)$ is given by evaluating the mode strength $b_l(k)$, as defined in Sect. 3.4.2, at discrete values of the wavenumber k corresponding to the frequency indices ν.

To cancel this dependence, we divide the eigenbeams by the mode strength (as in [21, 23, 34]), thus giving mode strength compensated eigenbeams, and the signal model is then written as

$$\begin{aligned} \widetilde{P}_{lm}(\nu) &= \left[\sqrt{4\pi} B_l(\nu)\right]^{-1} P_{lm}(\nu) \\ &= \widetilde{H}_{lm}(\nu)S(\nu) + \widetilde{V}_{lm,\text{c}}(\nu) + \widetilde{V}_{lm,\text{nc}}(\nu) \\ &= \widetilde{X}_{lm}(\nu) + \widetilde{V}_{lm,\text{c}}(\nu) + \widetilde{V}_{lm,\text{nc}}(\nu), \end{aligned} \tag{9.3}$$

where $\widetilde{P}_{lm}$, $\widetilde{H}_{lm}$, $\widetilde{X}_{lm}$, $\widetilde{V}_{lm,\text{c}}$ and $\widetilde{V}_{lm,\text{nc}}$ respectively denote the eigenbeams P_{lm}, H_{lm}, X_{lm}, $V_{lm,\text{c}}$ and $V_{lm,\text{nc}}$ after mode strength compensation.

As in Sect. 7.1, we choose as a reference a virtual omnidirectional microphone $\mathcal{M}_\text{ref}$ placed at the centre of the sphere, which is the origin of the spherical coordinate system. The signal $\widetilde{P}_{00}(\nu)$ is equal to the signal that would be received by the reference microphone [18, 21] $\mathcal{M}_\text{ref}$, if the sphere were not present, as shown in the Appendix of Chap. 5. Our aim is then to estimate the convolved source component $\widetilde{X}_{00}(k)$ of this signal, which we will hereafter refer to as the *desired signal*, using a tradeoff beamformer.

9.1.2 Tradeoff Beamformer

In this section, we present the tradeoff beamformer that we apply to the eigenbeams in order to achieve noise reduction. The tradeoff beamformer, introduced in Sect. 7.3.4, achieves a tradeoff between noise reduction and speech distortion. The weights of this beamformer are a function of second-order signal statistics, which can be estimated using the method that will be presented in Sect. 9.1.3.

For convenience, we rewrite the signal model (9.3) in vector notation, where the vectors all have length $N = (L+1)^2$, the total number of eigenbeams from order $l = 0$ to $l = L$:

$$\begin{aligned}\widetilde{\mathbf{p}}(\nu) &= \widetilde{\mathbf{h}}(\nu)S(\nu) + \widetilde{\mathbf{v}}_{\mathrm{c}}(\nu) + \widetilde{\mathbf{v}}_{\mathrm{nc}}(\nu) \\ &= \widetilde{\mathbf{x}}(\nu) + \widetilde{\mathbf{v}}_{\mathrm{c}}(\nu) + \widetilde{\mathbf{v}}_{\mathrm{nc}}(\nu) \\ &= \mathbf{d}(\nu)\widetilde{X}_{00}(\nu) + \widetilde{\mathbf{v}}(\nu), \end{aligned} \tag{9.4}$$

where $\mathbf{d}$ denotes a propagation vector of relative transfer functions (RTFs) given by

$$\mathbf{d}(\nu) = \left[1 \ \frac{\widetilde{H}_{1(-1)}(\nu)}{\widetilde{H}_{00}(\nu)} \ \frac{\widetilde{H}_{10}(\nu)}{\widetilde{H}_{00}(\nu)} \ \frac{\widetilde{H}_{11}(\nu)}{\widetilde{H}_{00}(\nu)} \ \cdots \ \frac{\widetilde{H}_{LL}(\nu)}{\widetilde{H}_{00}(\nu)}\right]^{\mathrm{T}},$$

$(\cdot)^{\mathrm{T}}$ denotes the vector transpose, the noisy signal vector $\widetilde{\mathbf{p}}$ is defined as

$$\widetilde{\mathbf{p}}(\nu) = \left[\widetilde{P}_{00}(\nu) \ \widetilde{P}_{1(-1)}(\nu) \ \widetilde{P}_{10}(\nu) \ \widetilde{P}_{11}(\nu) \ \cdots \ \widetilde{P}_{LL}(\nu)\right]^{\mathrm{T}},$$

and $\widetilde{\mathbf{x}}(\nu)$, $\widetilde{\mathbf{h}}(\nu)$, $\widetilde{\mathbf{v}}_{\mathrm{c}}(\nu)$ and $\widetilde{\mathbf{v}}_{\mathrm{nc}}(\nu)$ are defined similarly to $\widetilde{\mathbf{p}}(\nu)$. The coherent plus background noise signal vector $\widetilde{\mathbf{v}}$ is defined as $\widetilde{\mathbf{v}}(\nu) = \widetilde{\mathbf{v}}_{\mathrm{c}}(\nu) + \widetilde{\mathbf{v}}_{\mathrm{nc}}(\nu)$. We assume $H_{00}(\nu) \neq 0 \ \forall \nu$, such that $\mathbf{d}(\nu)$ is always defined.

The signals $X(\nu, \mathbf{r}_q)$, $V_{\mathrm{c}}(\nu, \mathbf{r}_q)$ and $V_{\mathrm{nc}}(\nu, \mathbf{r}_q)$ are mutually uncorrelated, and the SHT and division by the mode strength are linear operations, therefore $\widetilde{X}_{lm}(\nu)$, $\widetilde{V}_{lm,\mathrm{c}}(\nu)$ and $\widetilde{V}_{lm,\mathrm{nc}}(\nu)$ are also mutually uncorrelated. The PSD matrix $\mathbf{\Phi}_{\widetilde{\mathbf{p}}}$ of $\widetilde{\mathbf{p}}$ can therefore be decomposed as

$$\begin{aligned}\mathbf{\Phi}_{\widetilde{\mathbf{p}}}(\nu) &= \mathrm{E}\left\{\widetilde{\mathbf{p}}(\nu)\widetilde{\mathbf{p}}^{\mathrm{H}}(\nu)\right\} \\ &= \mathbf{\Phi}_{\widetilde{\mathbf{x}}}(\nu) + \mathbf{\Phi}_{\widetilde{\mathbf{v}}}(\nu) \\ &= \mathbf{\Phi}_{\widetilde{\mathbf{x}}}(\nu) + \mathbf{\Phi}_{\widetilde{\mathbf{v}}_{\mathrm{c}}}(\nu) + \mathbf{\Phi}_{\widetilde{\mathbf{v}}_{\mathrm{nc}}}(\nu), \end{aligned} \tag{9.5}$$

where $\mathrm{E}\{\cdot\}$ denotes mathematical expectation,

$$\begin{aligned}\mathbf{\Phi}_{\widetilde{\mathbf{x}}}(\nu) &= \mathrm{E}\left\{\widetilde{\mathbf{x}}(\nu)\widetilde{\mathbf{x}}^{\mathrm{H}}(\nu)\right\} = \phi_{\widetilde{X}_{00}}(\nu)\mathbf{d}(\nu)\mathbf{d}^{\mathrm{H}}(\nu), \\ \mathbf{\Phi}_{\widetilde{\mathbf{v}}}(\nu) &= \mathrm{E}\left\{\widetilde{\mathbf{v}}(\nu)\widetilde{\mathbf{v}}^{\mathrm{H}}(\nu)\right\} = \mathbf{\Phi}_{\widetilde{\mathbf{v}}_{\mathrm{c}}}(\nu) + \mathbf{\Phi}_{\widetilde{\mathbf{v}}_{\mathrm{nc}}}(\nu), \\ \mathbf{\Phi}_{\widetilde{\mathbf{v}}_{\mathrm{c}}}(\nu) &= \mathrm{E}\left\{\widetilde{\mathbf{v}}_{\mathrm{c}}(\nu)\widetilde{\mathbf{v}}_{\mathrm{c}}^{\mathrm{H}}(\nu)\right\} \text{ and} \\ \mathbf{\Phi}_{\widetilde{\mathbf{v}}_{\mathrm{nc}}}(\nu) &= \mathrm{E}\left\{\widetilde{\mathbf{v}}_{\mathrm{nc}}(\nu)\widetilde{\mathbf{v}}_{\mathrm{nc}}^{\mathrm{H}}(\nu)\right\}\end{aligned}$$

are respectively the PSD matrices of $\widetilde{\mathbf{x}}(\nu)$, $\widetilde{\mathbf{v}}(\nu)$, $\widetilde{\mathbf{v}}_{\mathrm{c}}(\nu)$, and $\widetilde{\mathbf{v}}_{\mathrm{nc}}(\nu)$, $\phi_{\widetilde{X}_{00}}(\nu) = \mathrm{E}\left\{|\widetilde{X}_{00}(\nu)|^2\right\}$ is the variance of $\widetilde{X}_{00}(\nu)$, and $(\cdot)^{\mathrm{H}}$ denotes the Hermitian transpose.

The output $Z(k)$ of our beamformer is obtained by applying a complex weight to each eigenbeam, and summing over all eigenbeams:

$$\begin{aligned} Z(\nu) &= \mathbf{w}^{\mathrm{H}}(\nu)\widetilde{\mathbf{p}}(\nu) \\ &= \mathbf{w}^{\mathrm{H}}(\nu)\widetilde{\mathbf{x}}(\nu) + \mathbf{w}^{\mathrm{H}}(\nu)\widetilde{\mathbf{v}}_{\mathrm{c}}(\nu) + \mathbf{w}^{\mathrm{H}}(\nu)\widetilde{\mathbf{v}}_{\mathrm{nc}}(\nu) \\ &= \widetilde{X}_{\mathrm{fd}}(\nu) + \widetilde{V}_{\mathrm{rc}}(\nu) + \widetilde{V}_{\mathrm{rnc}}(\nu), \end{aligned} \tag{9.6}$$

where $\widetilde{X}_{\mathrm{fd}}(\nu) = \mathbf{w}^{\mathrm{H}}(\nu)\widetilde{\mathbf{x}}(\nu) = \mathbf{w}^{\mathrm{H}}(\nu)\mathbf{d}(\nu)\widetilde{X}_{00}(\nu)$ is the filtered desired signal, $\widetilde{V}_{\mathrm{rc}}(\nu) = \mathbf{w}^{\mathrm{H}}(\nu)\widetilde{\mathbf{v}}_{\mathrm{c}}(\nu)$ is the residual coherent noise and $\widetilde{V}_{\mathrm{rnc}}(\nu) = \mathbf{w}^{\mathrm{H}}(\nu)\widetilde{\mathbf{v}}_{\mathrm{nc}}(\nu)$ is the residual background noise.

The tradeoff beamformer, introduced in Sect. 7.3.4, maximizes the noise reduction subject to a constraint on the speech distortion. Its weights are obtained by computing (7.50a) in the STFT domain, i.e.,

$$\mathbf{w}_{\mathrm{T},\mu}(\nu) = \frac{\phi_{\widetilde{X}_{00}}(\nu)\boldsymbol{\Phi}_{\widetilde{\mathbf{v}}}^{-1}(\nu)\mathbf{d}(\nu)}{\mu(\nu) + \phi_{\widetilde{X}_{00}}(\nu)\mathbf{d}^{\mathrm{H}}(\nu)\boldsymbol{\Phi}_{\widetilde{\mathbf{v}}}^{-1}(\nu)\mathbf{d}(\nu)}, \tag{9.7}$$

where $\phi_{\widetilde{V}_{00}}(\nu) = \mathrm{E}\left\{|\widetilde{V}_{00,\mathrm{c}}(\nu)|^2\right\} + \mathrm{E}\left\{|\widetilde{V}_{00,\mathrm{nc}}(\nu)|^2\right\}$ is the variance of $\widetilde{V}_{00}(\nu)$, and $\mu(\nu) \geq 0$ is the tradeoff parameter.

The higher the tradeoff parameter $\mu(\nu)$, the higher the noise reduction and the higher the speech distortion. The special case of $\mu = 0$ corresponds to a SHD minimum variance distortionless response (MVDR) beamformer, and $\mu = 1$ corresponds to a SHD Wiener filter. This parameter can also be signal-dependent, such that the tradeoff parameter is increased when only noise is present or likely to be present; for example, [31] proposed to control $\mu(\nu)$ using the SPP.

9.1.3 *Signal Statistics Estimation*

The computation of the tradeoff filter weights in (9.7) requires us to estimate both the noise PSD matrix $\boldsymbol{\Phi}_{\widetilde{\mathbf{v}}}$ and the RTF vector $\mathbf{d}$. The estimation of these second-order statistics has been the topic of extensive research; a common approach in the spatial domain has been to estimate them using the SPP [5, 6, 40, 43]. In this section, we review a SHD method that was proposed in Jarrett et al. [24] for this purpose, which is based on the DSPP.

Because speech is sparse in the STFT domain, a common assumption made by noise reduction algorithms is that in a sound field where multiple speech sources are present, only at most one of them is active in each time-frequency bin [1, 35]. This condition is known as (perfect) W-disjoint orthogonality. In [37], the authors showed that this is a good approximation if the STFT window parameters are chosen appropriately.

Consequently, for the purposes of the estimation of the second-order statistics, we adopt a model whereby at most one coherent source is active in each time-frequency bin: either the desired source, or an interfering source. While this model can break down in practice, for example when multiple interfering speakers are active (as in Sect. 9.1.6.2), typically only the desired source or the interfering sources are dominant in a single time-frequency bin, and the error in our model only causes a small amount of distortion in the filter output (see Sect. 9.1.6.3). It should be noted that this simplified model is only required for the statistics estimation; the tradeoff filter itself does not assume W-disjoint orthogonality.

Based on this model, if we denote the DOA of the active coherent source as $\Omega(\ell, \nu)$, we can formulate three hypotheses regarding the presence of speech:

$$\begin{aligned}
&\mathcal{H}_0(\ell, \nu) : \widetilde{\mathbf{p}}(\ell, \nu) = \widetilde{\mathbf{v}}_{\mathrm{nc}}(\ell, \nu) \\
&\qquad \text{indicating } \textit{speech absence}; \\
&\mathcal{H}_{1,\overline{\mathcal{R}}}(\ell, \nu) : \widetilde{\mathbf{p}}(\ell, \nu) = \widetilde{\mathbf{v}}_{\mathrm{c}}(\ell, \nu) + \widetilde{\mathbf{v}}_{\mathrm{nc}}(\ell, \nu) \\
&\qquad \text{indicating } \textit{interfering speech presence } [\Omega(\ell, \nu) \notin \mathcal{R}]; \\
&\mathcal{H}_{1,\mathcal{R}}(\ell, \nu) : \widetilde{\mathbf{p}}(\ell, \nu) = \widetilde{\mathbf{x}}(\ell, \nu) + \widetilde{\mathbf{v}}_{\mathrm{nc}}(\ell, \nu) \\
&\qquad \text{indicating } \textit{desired speech presence } [\Omega(\ell, \nu) \in \mathcal{R}].
\end{aligned}$$

In addition, we define the hypothesis $\mathcal{H}_1 = \mathcal{H}_{1,\overline{\mathcal{R}}} \cup \mathcal{H}_{1,\mathcal{R}}$, which indicates *speech presence* (desired or interfering). Under hypothesis $\mathcal{H}_{1,\mathcal{R}}$, the coherent source is located *inside* the region of interest $\mathcal{R}$, while under hypothesis $\mathcal{H}_{1,\overline{\mathcal{R}}}$, the coherent source is located *outside* $\mathcal{R}$.

9.1.3.1 Noise PSD Matrix Estimation

Based on these hypotheses, we can formulate a minimum mean square error estimate of the noise PSD matrix as[2]

$$\begin{aligned}
\mathrm{E}\left\{\widetilde{\mathbf{v}}\widetilde{\mathbf{v}}^{\mathrm{H}} | \widetilde{\mathbf{p}}\right\} = &\Pr\left[\mathcal{H}_0 \cup \mathcal{H}_{1,\overline{\mathcal{R}}} | \widetilde{\mathbf{p}}\right] \mathrm{E}\left\{\widetilde{\mathbf{v}}\widetilde{\mathbf{v}}^{\mathrm{H}} | \widetilde{\mathbf{p}}, \mathcal{H}_0 \cup \mathcal{H}_{1,\overline{\mathcal{R}}}\right\} \\
&+ \Pr\left[\mathcal{H}_{1,\mathcal{R}} | \widetilde{\mathbf{p}}\right] \mathrm{E}\left\{\widetilde{\mathbf{v}}\widetilde{\mathbf{v}}^{\mathrm{H}} | \widetilde{\mathbf{p}}, \mathcal{H}_{1,\mathcal{R}}\right\},
\end{aligned} \tag{9.8}$$

where $\Pr\left[\mathcal{H}_{1,\mathcal{R}} | \widetilde{\mathbf{p}}\right]$ is the a posteriori DSPP and $\Pr\left[\mathcal{H}_0 \cup \mathcal{H}_{1,\overline{\mathcal{R}}} | \widetilde{\mathbf{p}}\right] = 1 - \Pr\left[\mathcal{H}_{1,\mathcal{R}} | \widetilde{\mathbf{p}}\right]$ is the a posteriori desired speech absence probability. The estimation of the DSPP will be addressed in Sect. 9.1.4. Equation 9.8 is commonly approximated by recursively estimating the PSD matrix with an SPP-dependent smoothing factor [39, 43]; the PSD matrix estimate is then updated most rapidly in the absence of speech.

Close attention needs to be paid to the choice of the smoothing factor, which may be a compromise between the two following objectives: the noise PSD matrix should

[2]For brevity, the dependency of all quantities on the discrete time and frequency indices ℓ and ν is omitted where possible in the rest of Sect. 9.1.

be updated sufficiently slowly to avoid desired speech leaking into the estimate when the SPP is high, which would result in desired speech cancellation; the noise PSD matrix must also be updated sufficiently quickly to suppress any non-stationary noise present.

Because both the desired and coherent noise source signals consist of speech, which is non-stationary and has a similar spectral distribution regardless of the speaker (most of the energy is concentrated at low frequencies), we recursively estimate the PSD as

$$\hat{\mathbf{\Phi}}_{\widetilde{\mathbf{v}}}(\ell) = \alpha'_{\mathrm{v}}(\ell)\hat{\mathbf{\Phi}}_{\widetilde{\mathbf{v}}}(\ell-1) + \left[1-\alpha'_{\mathrm{v}}(\ell)\right]\widetilde{\mathbf{p}}(\ell)\widetilde{\mathbf{p}}^{\mathrm{H}}(\ell), \tag{9.9}$$

where

$$\alpha'_{\mathrm{v}} = \begin{cases} \alpha_{\mathrm{v}}, & \text{if } \Pr\left[\mathcal{H}_{1,\mathcal{R}}|\widetilde{\mathbf{p}}\right] < \Pr_{\mathrm{th}}; \\ 1, & \text{otherwise,} \end{cases} \tag{9.10}$$

and α_{v} is a smoothing factor between 0 and 1. As a result, the noise PSD matrix estimate is only updated if the DSPP is below a threshold $\Pr_{\mathrm{th}}$, which reduces the risk of desired speech cancellation, but also allows for rapid updates when speech is very unlikely to be present.

9.1.3.2 Relative Transfer Function Vector Estimation

As shown in Sect. 7.4.1, a minimum mean square error estimate of the RTF vector $\mathbf{d}$ is given by the first column of $\mathbf{\Phi}_{\widetilde{\mathbf{x}}}$ divided by its first element, $\phi_{\widetilde{X}_{00}}$, i.e.,

$$\widehat{\mathbf{d}} = \hat{\phi}_{\widetilde{X}_{00}}^{-1}\hat{\mathbf{\Phi}}_{\widetilde{\mathbf{x}}}\mathbf{i}_N, \tag{9.11}$$

where $\mathbf{i}_N = [1\ 0\ \cdots\ 0]^{\mathrm{T}}$ is a vector of length N, and $\hat{\mathbf{\Phi}}_{\widetilde{\mathbf{x}}}$ is an estimate of the PSD matrix of the convolved source signal $\widetilde{\mathbf{x}}$.

Unfortunately the convolved source signal cannot be directly observed, since the background noise is always present. We therefore first estimate the PSD matrix $\hat{\mathbf{\Phi}}_{\widetilde{\mathbf{x}}+\widetilde{\mathbf{v}}_{\mathrm{nc}}}$ of the desired speech plus background noise:

$$\hat{\mathbf{\Phi}}_{\widetilde{\mathbf{x}}+\widetilde{\mathbf{v}}_{\mathrm{nc}}}(\ell) = \alpha'_{\mathrm{xv_{nc}}}(\ell)\widetilde{\mathbf{p}}(\ell)\widetilde{\mathbf{p}}^{\mathrm{H}}(\ell) + [1-\alpha'_{\mathrm{xv_{nc}}}(\ell)]\hat{\mathbf{\Phi}}_{\widetilde{\mathbf{x}}+\widetilde{\mathbf{v}}_{\mathrm{nc}}}(\ell-1), \tag{9.12}$$

where $\alpha'_{\mathrm{xv_{nc}}} = \Pr\left[\mathcal{H}_{1,\mathcal{R}}|\widetilde{\mathbf{p}}\right](1-\alpha_{\mathrm{xv_{nc}}})$ and $\alpha_{\mathrm{xv_{nc}}}$ is a smoothing factor between 0 and 1. The convolved source PSD matrix can then be estimated as

$$\hat{\mathbf{\Phi}}_{\widetilde{\mathbf{x}}} = \hat{\mathbf{\Phi}}_{\widetilde{\mathbf{x}}+\widetilde{\mathbf{v}}_{\mathrm{nc}}} - \hat{\mathbf{\Phi}}_{\widetilde{\mathbf{v}}_{\mathrm{nc}}}. \tag{9.13}$$

In contrast to the noise PSD matrix, the convolved source PSD matrix and RTF vector estimates are most rapidly updated when the DSPP is high. An estimate $\hat{\mathbf{\Phi}}_{\widetilde{\mathbf{v}}_{\mathrm{nc}}}$

of the background noise PSD matrix can be obtained during an initial period where only background noise is present, assuming that the background noise is stationary. Alternatively, if the background noise is non-stationary, we can recursively estimate its PSD matrix using a smoothing factor that depends on $\Pr\left[\mathcal{H}_0|\widetilde{\mathbf{p}}\right]$, which is equal to $1 - \Pr\left[\mathcal{H}_1|\widetilde{\mathbf{p}}\right]$. The estimation of the a posteriori SPP $\Pr\left[\mathcal{H}_1|\widetilde{\mathbf{p}}\right]$ will be addressed in Sect. 9.1.4.1.

9.1.4 Desired Speech Presence Probability Estimation

Based on our hypotheses in Sect. 9.1.3, $\mathcal{H}_{1,\mathcal{R}} \cap \mathcal{H}_1 = \mathcal{H}_{1,\mathcal{R}}$; in other words, if desired speech is present, then speech must be present. We can therefore express the a posteriori DSPP $\Pr\left[\mathcal{H}_{1,\mathcal{R}}|\widetilde{\mathbf{p}}\right]$ as

$$\Pr\left[\mathcal{H}_{1,\mathcal{R}}|\widetilde{\mathbf{p}}\right] = \Pr\left[\mathcal{H}_{1,\mathcal{R}} \cap \mathcal{H}_1|\widetilde{\mathbf{p}}\right] \tag{9.14a}$$

$$= \Pr\left[\mathcal{H}_{1,\mathcal{R}}|\mathcal{H}_1, \widetilde{\mathbf{p}}\right] \cdot \Pr\left[\mathcal{H}_1|\widetilde{\mathbf{p}}\right]. \tag{9.14b}$$

An estimate of the a posteriori SPP $\Pr\left[\mathcal{H}_1|\widetilde{\mathbf{p}}\right]$ can be obtained using a Gaussian model–based multichannel SPP estimator proposed in Souden et al. [39]. When the noise signals contain coherent speech, the likelihood model in Souden et al. [39] does not reliably distinguish between desired and interfering sources, due to the non-stationarity of the signals and the inherent "chicken and egg" problem in the signal-based SPP estimation. The combination of the a posteriori SPP with the probability $\Pr\left[\mathcal{H}_{1,\mathcal{R}}|\mathcal{H}_1, \widetilde{\mathbf{p}}\right]$ enables us to differentiate between desired and interfering coherent sources.

Instead of directly using the noisy signal vector $\widetilde{\mathbf{p}}$ for inferring the probability $\Pr\left[\mathcal{H}_{1,\mathcal{R}}|\mathcal{H}_1, \widetilde{\mathbf{p}}\right]$, in Jarrett et al. [24], the authors proposed to compute instantaneous, narrowband DOA estimates $\hat{\Omega}$ from $\widetilde{\mathbf{p}}$, and to use these estimates to approximate the probability $\Pr\left[\mathcal{H}_{1,\mathcal{R}}|\mathcal{H}_1, \widetilde{\mathbf{p}}\right]$. The probability that is thereby computed is denoted as $\Pr[\mathcal{H}_{1,\mathcal{R}}|\mathcal{H}_1, \hat{\Omega}]$, and is referred to as a *DOA-based probability*. By using DOA estimates instead of the noisy signal vector $\widetilde{\mathbf{p}}$, the dimensionality of the problem is reduced. Furthermore, because the region of interest is defined in terms of DOAs, this approximation enables the use of intuitive and relatively accurate likelihood models, as will be seen in Sect. 9.1.4.2.

9.1.4.1 Multichannel Speech Presence Probability

The a posteriori SPP is estimated by assuming the desired speech, coherent noise and background noise signals can be modelled as complex multivariate Gaussian random variables. An estimate of the a posteriori multichannel SPP is then given by [39]

$$\Pr\left[\mathcal{H}_1|\widetilde{\mathbf{p}}\right] = \left\{1 + \frac{1-\varrho}{\varrho}(1+\xi)e^{-\frac{\beta}{1+\xi}}\right\}^{-1}, \tag{9.15}$$

where $\varrho = \Pr\left[\mathcal{H}_1\right]$ denotes the a priori SPP,

$$\beta = \widetilde{\mathbf{p}}^{\mathrm{H}}\hat{\mathbf{\Phi}}_{\widetilde{\mathbf{v}}_{\mathrm{nc}}}^{-1}\hat{\mathbf{\Phi}}_{\widetilde{\mathbf{r}}}\hat{\mathbf{\Phi}}_{\widetilde{\mathbf{v}}_{\mathrm{nc}}}^{-1}\widetilde{\mathbf{p}}, \tag{9.16}$$

and

$$\xi = \mathrm{tr}\left(\hat{\mathbf{\Phi}}_{\widetilde{\mathbf{v}}_{\mathrm{nc}}}^{-1}\hat{\mathbf{\Phi}}_{\widetilde{\mathbf{r}}}\right). \tag{9.17}$$

The PSD matrix $\hat{\mathbf{\Phi}}_{\widetilde{\mathbf{r}}}$ of the convolved source plus coherent noise signals is given by

$$\hat{\mathbf{\Phi}}_{\widetilde{\mathbf{r}}} = \hat{\mathbf{\Phi}}_{\widetilde{\mathbf{p}}} - \hat{\mathbf{\Phi}}_{\widetilde{\mathbf{v}}_{\mathrm{nc}}}. \tag{9.18}$$

The PSD matrix $\mathbf{\Phi}_{\widetilde{\mathbf{p}}}$ of the noisy signal vector $\widetilde{\mathbf{p}}$ is recursively estimated as

$$\hat{\mathbf{\Phi}}_{\widetilde{\mathbf{p}}}(\ell) = \alpha_{\mathrm{p}}\hat{\mathbf{\Phi}}_{\widetilde{\mathbf{p}}}(\ell-1) + (1-\alpha_{\mathrm{p}})\widetilde{\mathbf{p}}(\ell)\widetilde{\mathbf{p}}^{\mathrm{H}}(\ell), \tag{9.19}$$

where α_{p} is a smoothing factor between 0 and 1.

9.1.4.2 DOA-Based Probability

The DOA-based probability $\Pr[\mathcal{H}_{1,\mathcal{R}}|\mathcal{H}_1,\hat{\Omega}]$ can be determined from the instantaneous narrowband DOA estimates and the probability distribution function (PDF) $f(\hat{\Omega}|\mathcal{H}_1,\Omega)$ of the estimates $\hat{\Omega}$ obtained when a coherent source is present (hypothesis $\mathcal{H}_1$) with DOA $\Omega = (\theta,\phi)$.

More specifically, the DOA-based probability $\Pr[\mathcal{H}_{1,\mathcal{R}}|\mathcal{H}_1,\hat{\Omega}]$ is obtained by integrating the PDF $f(\Omega|\mathcal{H}_1,\hat{\Omega})$ over the region of interest $\mathcal{R}$, i.e.,

$$\Pr[\mathcal{H}_{1,\mathcal{R}}|\mathcal{H}_1,\hat{\Omega}] = \Pr[\Omega\in\mathcal{R}|\mathcal{H}_1,\hat{\Omega}] \tag{9.20a}$$

$$= \int_{\Omega\in\mathcal{R}} f(\Omega|\mathcal{H}_1,\hat{\Omega})\,\mathrm{d}\Omega \tag{9.20b}$$

$$= \int_{\Omega\in\mathcal{R}} \frac{f(\hat{\Omega}|\mathcal{H}_1,\Omega)f(\Omega|\mathcal{H}_1)}{f(\hat{\Omega}|\mathcal{H}_1)}\,\mathrm{d}\Omega, \tag{9.20c}$$

where $\mathrm{d}\Omega = \sin\theta\mathrm{d}\theta\mathrm{d}\phi$. Bayes' rule is used to obtain (9.20c) from (9.20b).

The marginal PDF $f(\Omega|\mathcal{H}_1)$ can be modelled using a priori knowledge of the desired source position. We assume that all DOAs Ω are equally likely, i.e., that $f(\Omega|\mathcal{H}_1)$ is uniform and hence equal to $\frac{1}{4\pi}$ for all values of Ω. The PDF $f(\hat{\Omega}|\mathcal{H}_1)$ can be computed by marginalizing over all possible DOAs Ω, i.e.,

$$f(\hat{\Omega}|\mathcal{H}_1) = \int_{\Omega \in \mathcal{S}^2} f(\hat{\Omega}|\mathcal{H}_1,\Omega) f(\Omega|\mathcal{H}_1)\,\mathrm{d}\Omega, \tag{9.21}$$

where the notation $\int_{\Omega \in \mathcal{S}^2} \mathrm{d}\Omega$ is used to denote compactly the solid angle $\int_{\phi=0}^{2\pi} \int_{\theta=0}^{\pi} \sin\theta \mathrm{d}\theta \mathrm{d}\phi$.

The PDF $f(\hat{\Omega}|\mathcal{H}_1,\Omega)$ is determined using a training phase, during which estimated DOA observations $\hat{\Omega}$ are collected under a number of specific conditions (such as the direct-to-reverberant ratio or signal-to-noise ratio). A parametric statistical model that fits the observations is then chosen, and its parameters can be estimated for every combination of training conditions.

The DOA estimates can be obtained using any of the narrowband DOA estimation algorithms presented in Sect. 5.1, as long as the training is performed with the chosen algorithm. We choose the pseudointensity vector method [19], described in Sect. 5.1.3. The unit vector that points in a direction Ω is denoted as $\mathbf{u}$. An estimate $\hat{\mathbf{u}}$ of $\mathbf{u}$ is then given by

$$\hat{\mathbf{u}}(\ell, \nu) = -\frac{\sum_{\ell'=\ell-\tau+1}^{\ell} \mathbf{I}(\ell', \nu)}{\| \sum_{\ell'=\ell-\tau+1}^{\ell} \mathbf{I}(\ell', \nu) \|}, \tag{9.22}$$

where $\mathbf{I}$ denotes the pseudointensity vector and $\| \cdot \|$ denotes the 2-norm. The moving average of the pseudointensity vectors over τ time frames gives the highest weight to the vectors with the highest norm, which are likely to be more reliable. The instantaneous, narrowband DOA estimate $\hat{\Omega}(\ell, \nu)$ is then given by the direction of the unit vector $\hat{\mathbf{u}}(\ell, \nu)$.

We use the von Mises–Fisher distribution [11], a probability distribution on the sphere,[3] to represent the DOA estimates thereby obtained. This distribution is rotationally symmetric about its mean direction; the deviation of the estimates from the mean direction is described by the concentration parameter κ. Due to the spherical symmetry of the microphone array, we can assume that the DOA estimates $\hat{\Omega}$ are unbiased for all values of Ω and that the mean direction is therefore equal to the true DOA Ω, i.e., $\mathrm{E}\left\{\hat{\Omega}|\mathcal{H}_1,\Omega\right\} = \Omega$. For the same reason, we can assume that the concentration parameter κ is independent of Ω, as long as the source and array are not near the room boundaries. A method for estimating the concentration parameter is set out in [41].

Using the von Mises–Fisher distribution to represent the DOA estimates obtained using the pseudointensity vector method, the PDF $f(\hat{\Omega}|\mathcal{H}_1,\Omega;\kappa)$ is given by [11, 26]

$$f(\hat{\Omega}|\mathcal{H}_1,\Omega;\kappa) = \frac{\kappa}{4\pi \sinh \kappa} e^{\kappa\, \mathbf{u}(\Omega)\cdot\hat{\mathbf{u}}(\hat{\Omega})} \tag{9.23a}$$

$$= \frac{\kappa}{2\pi\,(e^{\kappa} - e^{-\kappa})} e^{\kappa\, \mathbf{u}(\Omega)\cdot\hat{\mathbf{u}}(\hat{\Omega})}, \tag{9.23b}$$

[3] When the sphere is a 2-sphere (i.e., an ordinary sphere), as it is here, the von Mises–Fisher distribution is sometimes referred to simply as a *Fisher distribution*.

where $\mathbf{u}(\Omega) \cdot \hat{\mathbf{u}}(\hat{\Omega})$ denotes the scalar product of the vector $\mathbf{u}(\Omega)$, which points in a direction Ω, and the vector $\hat{\mathbf{u}}(\hat{\Omega})$, which points in a direction $\hat{\Omega}$. This scalar product is applied to the vectors in Cartesian coordinates, and is equal to the cosine of the angle between the true and estimated DOAs Ω and $\hat{\Omega}$. We refer to this angle as the *opening angle*. As κ decreases, the distribution of the opening angles becomes less concentrated around 0 and the distribution of the DOA estimates $\hat{\Omega}$ becomes less concentrated around the true DOA Ω.

As we have chosen a von Mises–Fisher distribution to represent the DOA estimates obtained using the pseudointensity vector method, and as we have assumed that $f(\Omega|\mathcal{H}_1)$ is uniform, that the concentration parameter κ is independent of Ω, and that the DOA estimates $\hat{\Omega}$ are unbiased, the DOA-based probability can advantageously be computed directly using (9.20b) rather than (9.20c). Indeed, using (9.21), the PDF $f(\hat{\Omega}|\mathcal{H}_1)$ can be computed as

$$\begin{aligned} f(\hat{\Omega}|\mathcal{H}_1) &= \int_{\Omega \in \mathcal{S}^2} f(\hat{\Omega}|\mathcal{H}_1, \Omega) f(\Omega|\mathcal{H}_1) \, \mathrm{d}\Omega && (9.24a) \\ &= \frac{1}{4\pi} \int_{\Omega \in \mathcal{S}^2} f(\hat{\Omega}|\mathcal{H}_1, \Omega) \, \mathrm{d}\Omega && (9.24b) \\ &= \frac{1}{4\pi} \int_{\Omega \in \mathcal{S}^2} \underbrace{\frac{\kappa}{2\pi \left(e^{\kappa} - e^{-\kappa}\right)} e^{\kappa \, \mathbf{u}(\Omega) \cdot \hat{\mathbf{u}}(\hat{\Omega})}}_{\star} \, \mathrm{d}\Omega. && (9.24c) \end{aligned}$$

By noticing that the expression marked with a $\star$ is also equal to the PDF of a random variable Ω that is distributed according to a von Mises–Fisher distribution with a concentration parameter κ and a mean direction $\hat{\Omega}$, we find that the integral of this expression over the sphere is equal to 1, and we obtain

$$f(\hat{\Omega}|\mathcal{H}_1) = \frac{1}{4\pi}. \tag{9.25}$$

The PDF $f(\hat{\Omega}|\mathcal{H}_1)$ is therefore uniform. As a result, using Bayes' rule, it follows that $f(\Omega|\mathcal{H}_1, \hat{\Omega}; \kappa) = f(\hat{\Omega}|\mathcal{H}_1, \Omega; \kappa)$, and $f(\Omega|\mathcal{H}_1, \hat{\Omega}; \kappa)$ can therefore be computed directly from (9.23).

9.1.5 *Algorithm Summary*

The estimation of the noise PSD matrix $\boldsymbol{\Phi}_{\tilde{\mathbf{v}}}$ and RTF vector $\mathbf{d}$ can be summarized as follows:

1. **Estimate the DOA-based probability** $\Pr[\mathcal{H}_{1,\mathcal{R}}(\ell, \nu)|\mathcal{H}_1(\ell, \nu), \hat{\Omega}(\ell, \nu)]$:
    a. Compute the pseudointensity vector $\mathbf{I}(\ell, \nu)$ by applying the method of Sect. 5.1.3 to the eigenbeams $\widetilde{P}_{lm}(\ell, \nu)$.

  b. Compute the unit vector $\hat{\mathbf{u}}(\ell, \nu)$ using the last τ pseudointensity vectors $\mathbf{I}(\ell - \tau+1, \nu), \ldots, \mathbf{I}(\ell - 1, \nu), \mathbf{I}(\ell, \nu)$ and (9.22).
  c. Compute the PDF $f(\hat{\Omega}|\mathcal{H}_1, \Omega; \kappa)$ using $\hat{\mathbf{u}}(\ell, \nu)$, the concentration parameter κ obtained from the training phase, and (9.23b).
  d. Estimate $\Pr[\mathcal{H}_{1,\mathcal{R}}(\ell, \nu)|\mathcal{H}_1(\ell, \nu), \hat{\Omega}(\ell, \nu)]$ using $f(\hat{\Omega}|\mathcal{H}_1, \Omega; \kappa)$ and (9.20b).

2. Update $\hat{\mathbf{\Phi}}_{\widetilde{\mathbf{p}}}(\ell, \nu)$ using (9.19).
3. Estimate $\mathbf{\Phi}_{\widetilde{\mathbf{r}}}(\ell, \nu)$ as $\hat{\mathbf{\Phi}}_{\widetilde{\mathbf{r}}}(\ell, \nu) = \hat{\mathbf{\Phi}}_{\widetilde{\mathbf{p}}}(\ell, \nu) - \hat{\mathbf{\Phi}}_{\widetilde{\mathbf{v}}_{\mathrm{nc}}}$. The PSD matrix $\hat{\mathbf{\Phi}}_{\widetilde{\mathbf{v}}_{\mathrm{nc}}}$ is estimated during initial frames where only background noise is present.
4. **Estimate the a posteriori multichannel SPP** $\Pr\left[\mathcal{H}_1(\ell, \nu)|\widetilde{\mathbf{p}}(\ell, \nu)\right]$ according to (9.15), (9.16) and (9.17), using $\hat{\mathbf{\Phi}}_{\widetilde{\mathbf{r}}}(\ell, \nu)$ and $\hat{\mathbf{\Phi}}_{\widetilde{\mathbf{v}}_{\mathrm{nc}}}$.
5. **Compute the a posteriori DSPP** $\Pr\left[\mathcal{H}_{1,\mathcal{R}}(\ell, \nu)|\widetilde{\mathbf{p}}(\ell, \nu)\right]$ as the product of the DOA-based probability $\Pr[\mathcal{H}_{1,\mathcal{R}}(\ell, \nu)|\mathcal{H}_1(\ell, \nu), \hat{\Omega}(\ell, \nu)]$ and the a posteriori SPP $\Pr\left[\mathcal{H}_1(\ell, \nu)|\widetilde{\mathbf{p}}(\ell, \nu)\right]$.
6. Update $\hat{\mathbf{\Phi}}_{\widetilde{\mathbf{v}}}(\ell, \nu)$ according to (9.9) using $\Pr\left[\mathcal{H}_{1,\mathcal{R}}(\ell, \nu)|\widetilde{\mathbf{p}}(\ell, \nu)\right]$.
7. Update $\hat{\mathbf{\Phi}}_{\widetilde{\mathbf{x}}+\widetilde{\mathbf{v}}_{\mathrm{nc}}}(\ell, \nu)$ according to (9.12) using $\Pr\left[\mathcal{H}_{1,\mathcal{R}}(\ell, \nu)|\widetilde{\mathbf{p}}(\ell, \nu)\right]$, and compute $\widehat{\mathbf{d}}(\ell, \nu)$ according to (9.11).

A block diagram of the complete informed noise reduction algorithm is shown in Fig. 9.2. The steps in the algorithm summary above are included in the blocks in the lower part of the figure.

9.1.6 Results

In this section, we provide some sample results to illustrate the performance of the informed noise reduction algorithm. A complete performance evaluation can be found in Jarrett et al. [24], where the performance is analysed in terms of the signal-to-noise ratio improvement, speech distortion index, coherent noise reduction factor, and background noise reduction factor.[4]

These results pertain to two aspects of the algorithm: the DSPP estimation method described in Sect. 9.1.4, which is used to estimate the second-order statistics in Sect. 9.1.3, and the tradeoff beamformer described in Sect. 9.1.2.

In previous work, an SPP-dependent tradeoff parameter μ has been used [31]. As in Jarrett et al. [24], we make the tradeoff parameter DSPP-dependent, such that

$$\mu(\ell, \nu) = \frac{1}{\eta \frac{1}{\mu'} + (1 - \eta)\Pr\left[\mathcal{H}_{1,\mathcal{R}}(\ell, \nu)|\widetilde{\mathbf{p}}(\ell, \nu)\right]}, \tag{9.26}$$

[4] A number of audio examples are also available at http://www.ee.ic.ac.uk/sap/sphdoa/.

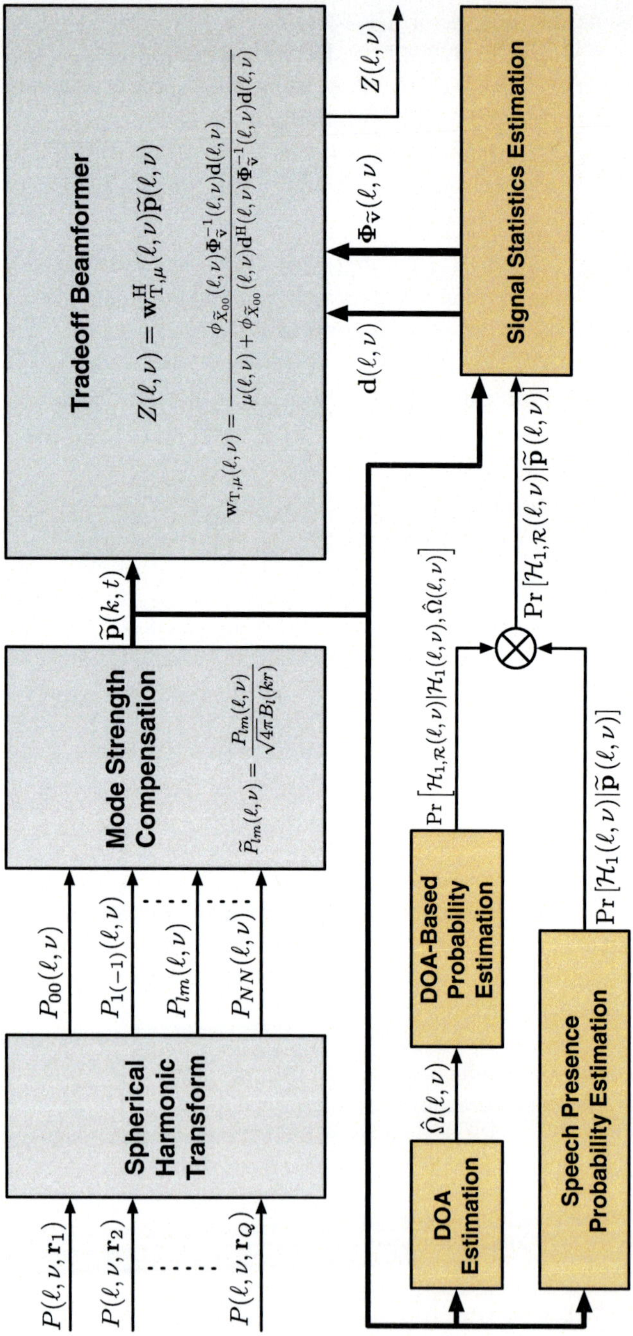

Fig. 9.2 Block diagram of the informed noise reduction algorithm, including the tradeoff beamformer described in Sect. 9.1.2 and the DOA-based statistics estimation algorithm presented in Sects. 9.1.3 and 9.1.4. The steps in the algorithm summary (Sect. 9.1.5) are included in the blocks in the lower part of the figure

where $0 \leq \eta \leq 1$ and $\mu > 0$. The smaller the value of η, the greater the influence of the DSPP $\Pr\left[\mathcal{H}_{1,\mathcal{R}}(\ell,\nu)|\widetilde{\mathbf{p}}(\ell,\nu)\right]$ on the tradeoff parameter μ. For the special case of $\eta = 1$, $\mu = \mu'$, and for $\eta = 0$, $\mu = \Pr\left[\mathcal{H}_{1,\mathcal{R}}(\ell,\nu)|\widetilde{\mathbf{p}}(\ell,\nu)\right]^{-1}$, $\forall \mu'$.

9.1.6.1 Experimental Setup

The results that follow were obtained by convolving clean speech signals with acoustic impulse responses (AIRs) that were measured in one of the laboratories at Fraunhofer IIS (Erlangen, Germany) [38]. The measurement room had a reverberation time T_{60} of approximately 330 ms, and dimensions of $7.5 \times 9.3 \times 4.2$ m. A rigid spherical array of radius 4.2 cm compromising $Q = 32$ microphones was placed approximately in the centre of the room. The desired source was placed at a distance of 1.8 m from the centre of an array, in a direction (95°, 175°) (inclination, azimuth). The first and second interfering sources were respectively placed 2.3 and 3.0 m away from the centre of the array, in directions (95°, 115°) and (40°, 0°).

The desired and interfering speech signals were taken from the EBU SQAM dataset [9]. Four 5 s segments were used, where the following speech sources were present: a desired source, a single interfering source, a desired source and a single interfering source, and a desired source and two interfering sources. Spatio-temporally white Gaussian noise was added to the pressure signals at each microphone such that a constant input signal-to-incoherent-noise ratio (iSINR) of 25 dB was obtained at $\mathcal{M}_{\text{ref}}$; as explained in Sect. 7.2.2, the incoherent noise power at $\mathcal{M}_{\text{ref}}$ is $Q|B_0(\nu)|^2$ times smaller than at the microphones [18]. The coherent noise level was set such that an input signal-to-coherent-noise ratio (iSCNR) of 0 dB was obtained at $\mathcal{M}_{\text{ref}}$, where the signal power was computed using only frames where *both* interfering talkers were active according to ITU-T Rec. P.56 [16]. Both the coherent and incoherent noise levels were chosen based on active speech levels, which were computed according to ITU-T Rec. P.56 [16].

The processing was performed at a sampling frequency of 8 kHz, with an STFT frame length of 512 samples (64 ms) and a 50 % overlap between frames. The STFT frames were zero-padded to 1024 samples (128 ms) before applying the fast Fourier transform in order to avoid circular convolution errors. We applied the beamformer to all eigenbeams up to order $L = 3$, but estimated the a posteriori SPP using only zero- and first-order eigenbeams, in order to reduce the computational complexity. We empirically chose the smoothing factors in (9.19), (9.10) and (9.12) as $\alpha_{\text{p}} = 0.8$, $\alpha_{\text{v}} = 0.7$ (with $\Pr_{\text{th}} = 0.01$) and $\alpha_{\text{xv}_{\text{nc}}} = 0.8$. We fixed the a priori SPP ϱ to 0.4, although the performance is not very sensitive to the choice of a priori SPP. Finally, we averaged the pseudointensity vectors over $\tau = 4$ time frames.

9.1.6.2 Desired Speech Presence Probability

The performance of the tradeoff beamformer is highly dependent on the accuracy of the estimated second-order statistics, which in turn depends on the accuracy of the DSPP estimation. We therefore first look at the DSPP estimation performance.

For the training phase required to estimate the concentration parameter κ of the von Mises–Fisher distribution, we used AIRs simulated with SMIRgen [17], an AIR simulator for spherical microphone arrays based on the algorithm presented in Chap. 4. We chose the same iSINR, reverberation time and source-array distance as in Sect. 9.1.6.1, thus yielding training conditions that were similar to the conditions where the tradeoff beamformer was applied. We numerically evaluated the integral in (9.20b) over the region of interest $\mathcal{R}$. The region of interest was centred around the desired source's true DOA:

$$\Omega = (\theta, \phi) \in \mathcal{R} \text{ if } \theta \in [80°, 110°] \text{ and } \phi \in [160°, 190°].$$

The narrowband DOA estimates can be biased by reverberation, especially when strong early reflections are present. To reduce this bias, and estimate a concentration parameter that is independent of the source-array position and DOA, we simulated 5 different source-array positions for each combination of the training parameters (iSINR, reverberation time, and source-array distance). We combined the results of these simulations to obtain a multimodal distribution, which we used to estimate the concentration parameter. It should be noted that because the array's directivity is frequency-dependent, so too is the concentration parameter. This can be seen in Fig. 9.3, where DOA estimates are shown for two frequencies; at low frequencies, the array has low directivity, and the DOA estimates are therefore less concentrated around the true DOA.

For illustration purposes, Fig. 9.4 shows some time-frequency plots of the opening angles between the estimated DOAs and the true DOA of the desired source (a), the DOA-based probability $\Pr[\mathcal{H}_{1,\mathcal{R}}|\mathcal{H}_1, \hat{\Omega}]$ (b), the a posteriori multichannel SPP

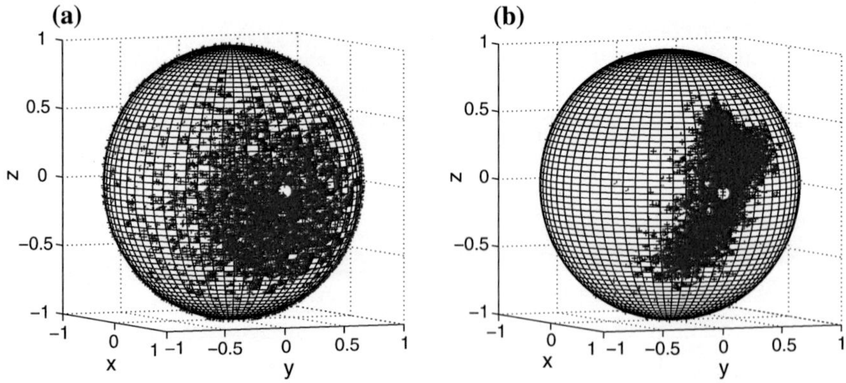

Fig. 9.3 DOA estimates obtained with 5 different source-array positions and a single true DOA (marked with a *white dot*), at **a** 100 Hz and **b** 2.2 kHz

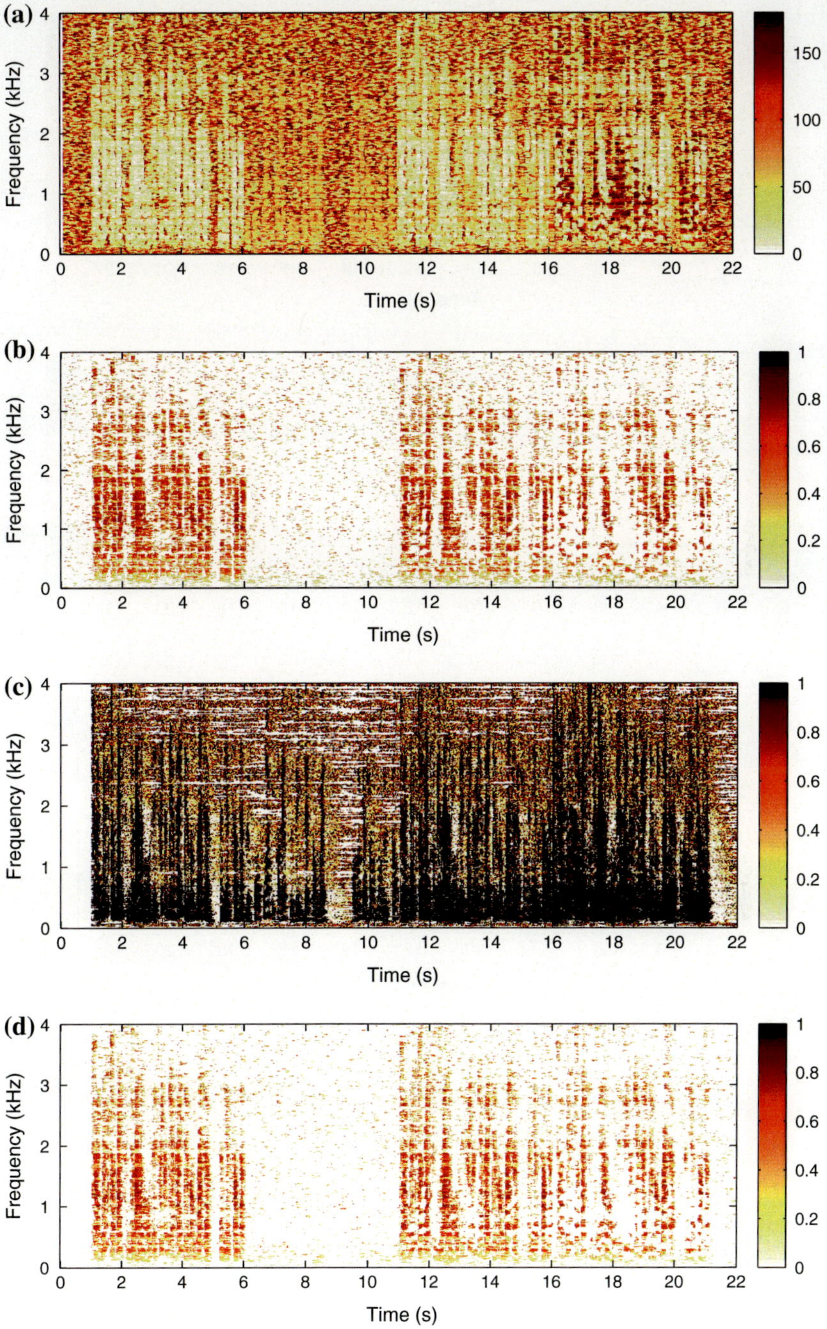

Fig. 9.4 Illustrative plots of the **a** opening angles (in degrees), **b** DOA-based probability $\Pr[\mathcal{H}_{1,\mathcal{R}}|\mathcal{H}_1, \hat{\Omega}]$, **c** a posteriori multichannel SPP $\Pr[\mathcal{H}_1|\tilde{\mathbf{p}}]$, and **d** a posteriori DSPP $\Pr[\mathcal{H}_{1,\mathcal{R}}|\tilde{\mathbf{p}}]$, as a function of time and frequency

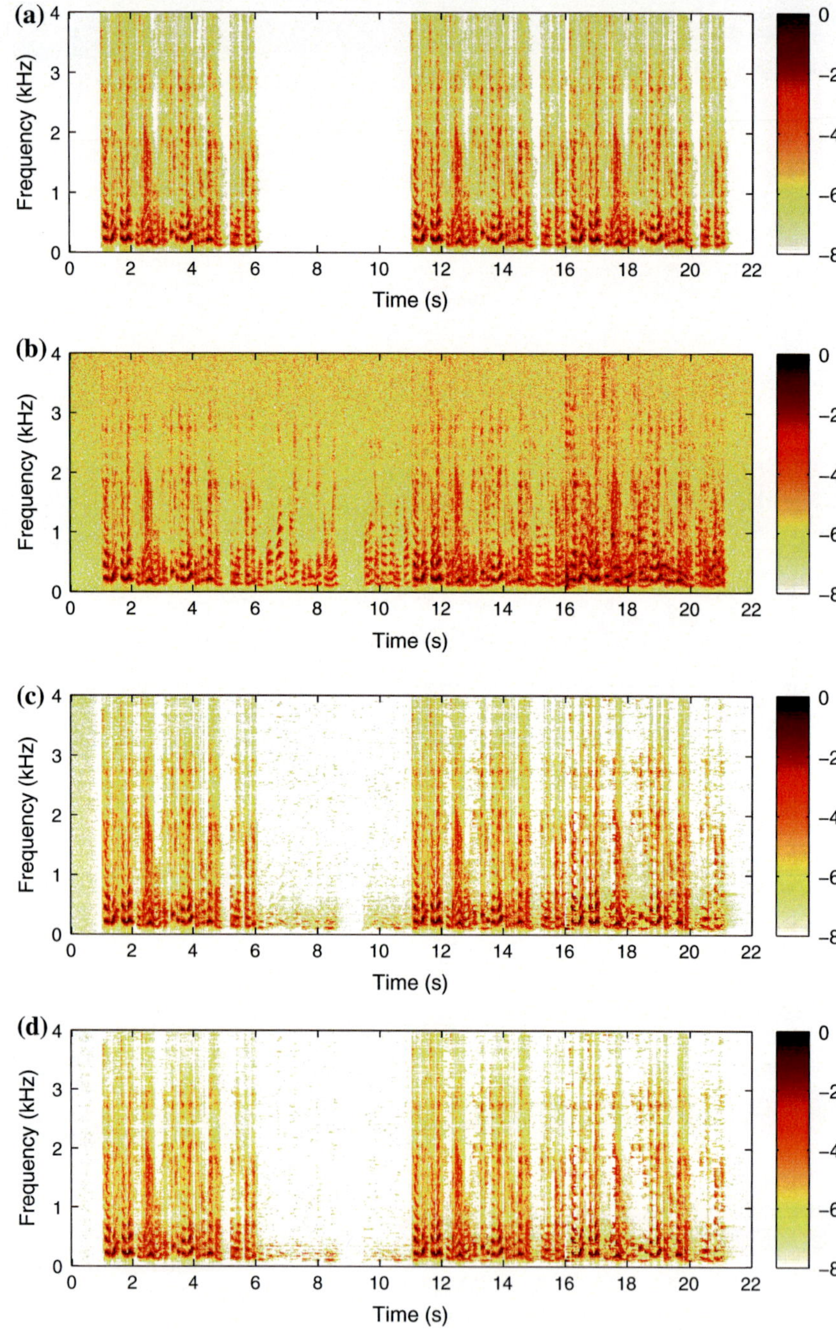

Fig. 9.5 Spectrograms of the **a** desired speech signal $\widetilde{X}_{00}$, **b** received signal $\widetilde{P}_{00} = \widetilde{X}_{00} + \widetilde{V}_{00,\mathrm{c}} + \widetilde{V}_{00,\mathrm{nc}}$, beamformer output Z for **c** $\eta = 1$, $\mu' \to 0$ and **d** $\eta = 0.25$, $\mu' = 3$

$\Pr\left[\mathcal{H}_1|\widetilde{\mathbf{p}}\right]$ (c), and the a posteriori DSPP $\Pr\left[\mathcal{H}_{1,\mathcal{R}}|\widetilde{\mathbf{p}}\right]$ (d). We see that the DOA-based probability efficiently distinguishes between desired and interfering sources; in time-frequency bins where the desired source is not present, between 6 and 11 s for instance, the DOA-based probability is low. Low values of the DOA-based probability result in a low a posteriori DSPP, which allows us to confidently update the noise PSD matrix $\mathbf{\Phi}_{\widetilde{\mathbf{v}}}$ without the risk of significant desired speech cancellation. As expected, in isolation the a posteriori SPP in (c) only allows us to detect coherent sources, whether they be desired or interfering.

9.1.6.3 Tradeoff Beamformer

In Fig. 9.5, we plot a number of sample spectrograms in order to show the effect of the informed noise reduction algorithm. Figure 9.5a, b respectively show the desired speech signal $\widetilde{X}_{00}(\ell, \nu)$ and the noisy signal $\widetilde{P}_{00}(\ell, \nu)$ measured at the reference microphone $\mathcal{M}_{\text{ref}}$. The effect of the sensor noise is most clearly seen when none of the coherent sources are active, for example, between 0 and 1 s. The sensor noise has higher power at high frequencies due to the fact that the signal $\widetilde{P}_{00}(\ell, \nu)$ is obtained by dividing the zero-order eigenbeam $P_{00}(\ell, \nu)$ by the mode strength $B_0(\nu)$.

The spectrograms of the tradeoff beamformer output for two sets of tradeoff parameters are shown in Fig. 9.5c, d. In Fig. 9.5c, $\mu' \to 0$, and hence $\mu = 0$; in this case the tradeoff beamformer is equivalent to the SHD MVDR beamformer. The interfering source has been efficiently suppressed, although a small amount of interfering speech remains at low frequencies, where the interfering speech has highest energy. By making the tradeoff parameter DSPP-dependent, as in Fig. 9.5d where $\eta = 0.25$ and $\mu' = 3$ (resulting in a tradeoff parameter μ in the range 1.2–12), a larger amount of noise reduction is achieved, at the expense of a slight increase in speech distortion, visible at high frequencies between 5 and 6 s, for example.

9.2 Dereverberation Using Signal-to-Diffuse Ratio Estimates

Reverberation is the phenomenon whereby the sound waves produced by an acoustic source in an enclosed space are reflected by the surrounding walls. Reverberation is known to be potentially detrimental to speech communication [30]. In particular, in the presence of high levels of reverberation, the intelligibility of speech may be degraded. Dereverberation techniques seek to mitigate the effects of reverberation; both single and multichannel techniques have been proposed for this purpose, detailed in Habets [12], Naylor and Gaubitch [30] and the references therein. This is a challenging problem, in large part because reverberation is highly time-varying and, unlike noise, cannot be observed in periods where the desired sources are inactive.

Dereverberation is the task of removing the effects of reverberation and is commonly approached using microphone arrays, where spatial filters are applied to the signals captured by each of the array's microphones to attenuate both the levels of reverberation and of ambient noise. These spatial filters can broadly be divided into two categories: signal-independent and signal-dependent filters, respectively explored in Chaps. 6 and 7. Reverberation is commonly modelled as spatially diffuse noise; hence, the tradeoff between white noise gain (WNG) and directivity discussed in Sect. 6.3 implies a tradeoff between WNG and dereverberation. In addition, it has been shown that there is a tradeoff between the noise reduction and dereverberation achieved using a MVDR filter [14]. A notable approach to multichannel dereverberation combines a (signal-independent) MVDR filter with a single-channel Wiener filter (a post-filter) [27].

While noise reduction in the SHD has received considerable attention, the topic of dereverberation in the SHD [20, 32–34, 46] is only in its infancy. In [34], the authors propose a method for noise reduction and dereverberation based on a linearly constrained minimum variance (LCMV) with spatial nulls in the direction of the reflections. However, in practice the DOAs of the reflections can be difficult to estimate. In addition, the specular reflections whose DOA can be estimated are likely to be early reflections, which contribute positively to intelligibility [2, 25], while the reflections that reduce intelligibility are more likely to be diffuse [29, 42].

In this section, we apply the concept of informed array processing to the problem of dereverberation in the SHD. We model the signal captured by a spherical microphone array as the sum of a direct signal, a diffuse signal that models reverberation, and a noise signal. All three signals are assumed to be mutually uncorrelated. We present an approach proposed by Braun et al. [3], where an optimal filter is derived that minimizes the mean square error (MSE) between the direct signal received at a reference microphone and the estimated direct signal. The weights of the resulting Wiener filter depend on the DOA of the direct signal and the PSD matrix of the interference signal, which is composed of both the diffuse and noise signals. As these two signals are uncorrelated, the PSD matrix of the interference signal is given by the sum of the PSD matrices of the diffuse and noise signals. The PSD matrix of the diffuse signal depends only on the power of the diffuse signal at the reference microphone, and can be estimated based on instantaneous SDR estimates (see Sect. 5.2). This approach is *informed* in the sense that information about the SDR is incorporated into the design of the spatial filter.

9.2.1 Problem Formulation

Signal Model

We consider a frequency domain signal model in which a spherical microphone array captures a direct signal X, a diffuse signal F and a noise signal V. The spatial domain signal received at Q microphone positions $\mathbf{r}_q = (r, \Omega_q) = (r, \theta_q, \phi_q)$,

$q \in \{1, \ldots, Q\}$ (in spherical coordinates, where θ denotes the inclination and ϕ denotes the azimuth) can then be expressed as[5]

$$P(\nu, \mathbf{r}_q) = X(\nu, \mathbf{r}_q) + F(\nu, \mathbf{r}_q) + V(\nu, \mathbf{r}_q) \tag{9.27}$$

We assume that the signals X, F and V are mutually uncorrelated. The diffuse signal F models reverberation that we wish to suppress.

When using spherical microphone arrays, it is convenient to work in the SHD [28, 36], instead of the spatial domain. In this chapter, we assume error-free spatial sampling, and refer the reader to Chap. 3 for information on spatial sampling and aliasing. By applying the complex SHT to the signal model in (7.1), we obtain the SHD signal model

$$P_{lm}(\nu) = X_{lm}(\nu) + F_{lm}(\nu) + V_{lm}(\nu) \tag{9.28}$$

where $P_{lm}(\nu)$, $X_{lm}(\nu)$, $F_{lm}(\nu)$ and $V_{lm}(\nu)$ are respectively the spherical harmonic transforms of the signals $P(\nu, \mathbf{r}_q)$, $X(\nu, \mathbf{r}_q)$, $F(\nu, \mathbf{r}_q)$ and $V(\nu, \mathbf{r}_q)$, as defined in (3.6), and are referred to as *eigenbeams* to reflect the fact that the spherical harmonics are eigensolutions of the wave equation in spherical coordinates [44]. The order and degree of the spherical harmonics are respectively denoted as l and m.

The eigenbeams $P_{lm}(\nu)$, $X_{lm}(\nu)$, $F_{lm}(\nu)$ and $V_{lm}(\nu)$ are a function of the frequency-dependent mode strength $b_l(\nu)$, which is in turn a function of the array properties (radius, microphone type, configuration). Mode strength expressions for two common types of arrays, the open and rigid arrays with omnidirectional microphones, are given in Sect. 3.4.2. To cancel this dependence, we divide the eigenbeams by the mode strength (as in [34]), thus giving mode strength compensated eigenbeams, and the signal model is then written as

$$\widetilde{P}_{lm}(\nu) = \left[\sqrt{4\pi} b_l(\nu)\right]^{-1} P_{lm}(\nu) \tag{9.29a}$$

$$= \widetilde{X}_{lm}(\nu) + \widetilde{F}_{lm}(\nu) + \widetilde{V}_{lm}(\nu) \tag{9.29b}$$

where $\widetilde{P}_{lm}(\nu)$, $\widetilde{X}_{lm}(\nu)$, $\widetilde{F}_{lm}(\nu)$ and $\widetilde{V}_{lm}(\nu)$ respectively denote the eigenbeams $P_{lm}(\nu)$, $X_{lm}(\nu)$, $F_{lm}(\nu)$ and $V_{lm}(\nu)$ after mode strength compensation.

Beamforming in the Spherical Harmonic Domain

As shown in the Appendix of Chap. 5, the eigenbeam $\widetilde{P}_{00}(\nu)$ is equal to the signal that would be received at an omnidirectional reference microphone $\mathcal{M}_{\text{ref}}$ positioned at the centre of the sphere, if the sphere were not present. Our objective is to derive a spatial filter or beamformer to estimate the direct component of the eigenbeam $\widetilde{P}_{00}(\nu)$, namely, $\widetilde{X}_{00}(\nu)$, which we refer to as the *desired signal*.

[5] The dependency on time is omitted for brevity. In practice, the signals acquired using a spherical microphone array are usually processed in the short-time Fourier transform domain, as explained in Sect. 3.1, where the discrete frequency index is denoted by ν.

For convenience, we rewrite the signal model (9.29) in vector notation, where the vectors all have length $N = (L+1)^2$, the total number of eigenbeams from order $l = 0$ to $l = L$:

$$\widetilde{\mathbf{p}}(\nu) = \widetilde{\mathbf{x}}(\nu) + \widetilde{\mathbf{f}}(\nu) + \widetilde{\mathbf{v}}(\nu) \tag{9.30}$$

where

$$\widetilde{\mathbf{p}}(\nu) = \left[\widetilde{P}_{00}(\nu)\ \widetilde{P}_{1(-1)}(\nu)\ \widetilde{P}_{10}(\nu)\ \widetilde{P}_{11}(\nu)\ \widetilde{P}_{2(-2)}(\nu) \cdots \widetilde{P}_{LL}(\nu)\right]^{\mathrm{T}}, \tag{9.31}$$

$(\cdot)^{\mathrm{T}}$ denotes the vector transpose, and $\widetilde{\mathbf{x}}(\nu)$, $\widetilde{\mathbf{f}}(\nu)$ and $\widetilde{\mathbf{v}}(\nu)$ are defined similarly to $\widetilde{\mathbf{p}}(\nu)$. We also define an interference signal vector $\widetilde{\mathbf{u}}(\nu) = \widetilde{\mathbf{f}}(\nu) + \widetilde{\mathbf{v}}(\nu)$, which contains the diffuse and noise signals that we wish to suppress.

Assuming that the direct signal $\widetilde{\mathbf{x}}$ is composed of a single plane wave with DOA Ω_{dir}, the signal vector $\widetilde{\mathbf{x}}(\nu)$ can be written in terms of the desired signal as

$$\widetilde{\mathbf{x}}(\nu) = \mathbf{d}_{\mathrm{dir}}(\nu)\widetilde{X}_{00}(\nu), \tag{9.32}$$

where, using (3.22a), we find that

$$\mathbf{d}_{\mathrm{dir}}(\nu) = \left[1\ \frac{Y^*_{1(-1)}(\Omega_{\mathrm{dir}})}{Y^*_{00}(\Omega_{\mathrm{dir}})}\ \frac{Y^*_{10}(\Omega_{\mathrm{dir}})}{Y^*_{00}(\Omega_{\mathrm{dir}})}\ \frac{Y^*_{11}(\Omega_{\mathrm{dir}})}{Y^*_{00}(\Omega_{\mathrm{dir}})} \cdots \frac{Y^*_{LL}(\Omega_{\mathrm{dir}})}{Y^*_{00}(\Omega_{\mathrm{dir}})}\right]^{\mathrm{T}} \tag{9.33}$$

for all frequencies ν, where Y_{lm} denotes the complex spherical harmonic[6] of order l and degree m, as defined in (2.14). The vector $\mathbf{d}_{\mathrm{dir}}(\nu)$ is referred to as a *steering vector*. As $X(\nu, \mathbf{r}_q)$, $F(\nu, \mathbf{r}_q)$ and $V(\nu, \mathbf{r}_q)$ are mutually uncorrelated, and the SHT and division by the mode strength are linear operations, $\widetilde{X}_{lm}(\nu)$, $\widetilde{F}_{lm}(\nu)$ and $\widetilde{V}_{lm}(\nu)$ are also mutually uncorrelated. The PSD matrix $\boldsymbol{\Phi}_{\widetilde{\mathbf{p}}}$ of $\widetilde{\mathbf{p}}$ can therefore be expressed as

$$\begin{aligned}\boldsymbol{\Phi}_{\widetilde{\mathbf{p}}}(\nu) &= \mathrm{E}\left\{\widetilde{\mathbf{p}}(\nu)\widetilde{\mathbf{p}}^{\mathrm{H}}(\nu)\right\} \\ &= \boldsymbol{\Phi}_{\widetilde{\mathbf{x}}}(\nu) + \boldsymbol{\Phi}_{\widetilde{\mathbf{f}}}(\nu) + \boldsymbol{\Phi}_{\widetilde{\mathbf{v}}}(\nu) \\ &= \boldsymbol{\Phi}_{\widetilde{\mathbf{x}}}(\nu) + \boldsymbol{\Phi}_{\widetilde{\mathbf{u}}}(\nu),\end{aligned} \tag{9.34}$$

where $\mathrm{E}\{\cdot\}$ denotes mathematical expectation and

$$\begin{aligned}\boldsymbol{\Phi}_{\widetilde{\mathbf{x}}}(\nu) &= \mathrm{E}\left\{\widetilde{\mathbf{x}}(\nu)\widetilde{\mathbf{x}}^{\mathrm{H}}(\nu)\right\} = \phi_{\widetilde{X}_{00}}(\nu)\mathbf{d}_{\mathrm{dir}}\mathbf{d}_{\mathrm{dir}}^{\mathrm{H}}, \\ \boldsymbol{\Phi}_{\widetilde{\mathbf{f}}}(\nu) &= \mathrm{E}\left\{\widetilde{\mathbf{f}}(\nu)\widetilde{\mathbf{f}}^{\mathrm{H}}(\nu)\right\} \text{ and} \\ \boldsymbol{\Phi}_{\widetilde{\mathbf{v}}}(\nu) &= \mathrm{E}\left\{\widetilde{\mathbf{v}}(\nu)\widetilde{\mathbf{v}}^{\mathrm{H}}(\nu)\right\}\end{aligned}$$

[6]If the real SHT is applied instead of the complex SHT, the complex spherical harmonics Y_{lm} used throughout this chapter should be replaced with the real spherical harmonics R_{lm}, as defined in Sect. 3.3.

are respectively the PSD matrices of $\widetilde{\mathbf{x}}(\nu)$, $\widetilde{\mathbf{f}}(\nu)$ and $\widetilde{\mathbf{v}}(\nu)$, $(\cdot)^{\mathrm{H}}$ denotes the Hermitian transpose, and $\phi_{\widetilde{X}_{00}}(\nu) = \mathrm{E}\left\{\left|\widetilde{X}_{00}(\nu)\right|^2\right\}$ is the variance of $\widetilde{X}_{00}(\nu)$.

As in Chap. 7, the output $Z(\nu)$ of our beamformer is obtained by applying a complex weight to each eigenbeam and summing over all eigenbeams:

$$\begin{aligned} Z(\nu) &= \mathbf{w}^{\mathrm{H}}(\nu)\widetilde{\mathbf{p}}(\nu) \\ &= \mathbf{w}^{\mathrm{H}}(\nu)\widetilde{\mathbf{x}}(\nu) + \mathbf{w}^{\mathrm{H}}(\nu)\widetilde{\mathbf{f}}(\nu) + \mathbf{w}^{\mathrm{H}}(\nu)\widetilde{\mathbf{v}}(\nu) \\ &= \widetilde{X}_{\mathrm{f}}(\nu) + \widetilde{F}_{\mathrm{r}}(\nu) + \widetilde{V}_{\mathrm{r}}(\nu), \end{aligned} \tag{9.35}$$

where $\widetilde{X}_{\mathrm{f}}(\nu) = \mathbf{w}^{\mathrm{H}}(\nu)\widetilde{\mathbf{x}}(\nu) = \mathbf{w}^{\mathrm{H}}(\nu)\mathbf{d}_{\mathrm{dir}}\widetilde{X}_{00}(\nu)$ is the filtered desired signal, $\widetilde{F}_{\mathrm{r}}(\nu) = \mathbf{w}^{\mathrm{H}}(\nu)\widetilde{\mathbf{f}}(\nu)$ is the residual diffuse signal and $\widetilde{V}_{\mathrm{r}}(\nu) = \mathbf{w}^{\mathrm{H}}(\nu)\widetilde{\mathbf{v}}(\nu)$ is the residual noise.

In the next section, we derive the weights of a spatial filter that estimates the desired signal $\widetilde{X}_{00}(\nu)$.

9.2.2 *Informed Filter for Dereverberation*

Filter Weights

The MSE between the filter output $Z(\nu)$ and the desired signal $\widetilde{X}_{00}(\nu)$ is given by

$$\begin{aligned} J\left[\mathbf{w}(\nu)\right] &= \mathrm{E}\left\{\left|Z(\nu) - \widetilde{X}_{00}(\nu)\right|^2\right\} \\ &= \mathrm{E}\left\{\left|\mathbf{w}^{\mathrm{H}}(\nu)\left[\mathbf{d}_{\mathrm{dir}}\widetilde{X}_{00}(\nu) + \widetilde{\mathbf{u}}(\nu)\right] - \widetilde{X}_{00}(\nu)\right|^2\right\}. \end{aligned} \tag{9.36}$$

The filter that minimizes the MSE is known as a multichannel Wiener filter, and is presented in Sect. 7.3.2. Its weights are given by

$$\mathbf{w}_{\mathrm{MWF}}(\nu) = \frac{\phi_{\widetilde{X}_{00}}(\nu)\boldsymbol{\Phi}_{\widetilde{\mathbf{u}}}^{-1}(\nu)\mathbf{d}_{\mathrm{dir}}}{1 + \phi_{\widetilde{X}_{00}}(\nu)\mathbf{d}_{\mathrm{dir}}^{\mathrm{H}}\boldsymbol{\Phi}_{\widetilde{\mathbf{u}}}^{-1}(\nu)\mathbf{d}_{\mathrm{dir}}}. \tag{9.37}$$

It can sometimes be advantageous to separate the weights in (9.37) into a MVDR filter and a single-channel Wiener filter [3, 21]:

$$\mathbf{w}_{\mathrm{MWF}}(\nu) = \underbrace{\frac{\boldsymbol{\Phi}_{\widetilde{\mathbf{u}}}^{-1}(\nu)\mathbf{d}_{\mathrm{dir}}}{\mathbf{d}_{\mathrm{dir}}^{\mathrm{H}}\boldsymbol{\Phi}_{\widetilde{\mathbf{u}}}^{-1}(\nu)\mathbf{d}_{\mathrm{dir}}}}_{\mathbf{w}_{\mathrm{MVDR}}(\nu)} \cdot \underbrace{\frac{\phi_{\widetilde{X}_{\mathrm{f,MVDR}}}(\nu)}{\phi_{\widetilde{X}_{\mathrm{f,MVDR}}}(\nu) + \phi_{\widetilde{U}_{\mathrm{r,MVDR}}}(\nu)}}_{W_{\mathrm{WF}}(\nu)}, \tag{9.38}$$

where

$$\begin{aligned} \phi_{\widetilde{X}_{\mathrm{f,MVDR}}}(\nu) &= \mathrm{E}\left\{|\mathbf{w}_{\mathrm{MVDR}}^{\mathrm{H}}(\nu)\widetilde{\mathbf{x}}(\nu)|^2\right\} \\ &= \mathbf{w}_{\mathrm{MVDR}}^{\mathrm{H}}(\nu)\boldsymbol{\Phi}_{\widetilde{\mathbf{x}}}(\nu)\mathbf{w}_{\mathrm{MVDR}}(\nu) && (9.39\mathrm{a}) \\ &= \mathbf{w}_{\mathrm{MVDR}}^{\mathrm{H}}(\nu)\left[\boldsymbol{\Phi}_{\widetilde{\mathbf{p}}}(\nu) - \boldsymbol{\Phi}_{\widetilde{\mathbf{u}}}(\nu)\right]\mathbf{w}_{\mathrm{MVDR}}(\nu) && (9.39\mathrm{b}) \end{aligned}$$

denotes the variance of the direct signal component of the output of the MVDR filter, and

$$\begin{aligned}\phi_{\widetilde{U}_{\mathrm{r,MVDR}}}(\nu) &= \mathrm{E}\left\{|\mathbf{w}_{\mathrm{MVDR}}^{\mathrm{H}}(\nu)\widetilde{\mathbf{u}}(\nu)|^2\right\} \\ &= \mathbf{w}_{\mathrm{MVDR}}^{\mathrm{H}}(\nu)\mathbf{\Phi}_{\widetilde{\mathbf{u}}}(\nu)\mathbf{w}_{\mathrm{MVDR}}(\nu) \qquad (9.40) \\ &= \left[\mathbf{d}_{\mathrm{dir}}^{\mathrm{H}}\mathbf{\Phi}_{\widetilde{\mathbf{u}}}^{-1}(\nu)\mathbf{d}_{\mathrm{dir}}\right]^{-1} \qquad (9.41)\end{aligned}$$

denotes the variance of the residual interference at the output of the MVDR filter. This form of the multichannel Wiener filter is advantageous as it provides better control over the speech distortion.

It should be noted that the weights of the signal-dependent MVDR filter in (9.38) depend on the statistics of both the diffuse and noise signals; the noise signal is not required to be diffuse. In contrast, the MVDR filter in [27] is signal-independent: because a diffuse noise field is assumed, the filter weights only depend on the coherence matrix of this field.

Signal Statistics Estimation

In order to compute the weights of the multichannel Wiener filter $\mathbf{w}_{\mathrm{MWF}}(\nu)$, we must estimate the PSD matrix $\mathbf{\Phi}_{\widetilde{\mathbf{u}}}(\nu)$ of the interference signals

$$\mathbf{\Phi}_{\widetilde{\mathbf{u}}}(\nu) = \mathbf{\Phi}_{\widetilde{\mathbf{f}}}(\nu) + \mathbf{\Phi}_{\widetilde{\mathbf{v}}}(\nu). \qquad (9.42)$$

We assume that the noise is stationary, and its PSD matrix $\mathbf{\Phi}_{\widetilde{\mathbf{v}}}(\nu)$ can therefore be estimated when the direct and diffuse signals are inactive. On the other hand, the diffuse signal used to model reverberation is highly *non-stationary*, and its PSD matrix $\mathbf{\Phi}_{\widetilde{\mathbf{f}}}(\nu)$ therefore must be continuously estimated.

The diffuse PSD matrix can be expressed as a function of the variance of $\widetilde{F}_{00}(\nu)$ and the diffuse coherence matrix $\mathbf{\Gamma}_{\widetilde{\mathbf{f}}}(\nu)$ such that

$$\mathbf{\Phi}_{\widetilde{\mathbf{f}}}(\nu) = \phi_{\widetilde{F}_{00}}(\nu)\mathbf{\Gamma}_{\widetilde{\mathbf{f}}}(\nu), \qquad (9.43)$$

where $\phi_{\widetilde{F}_{00}}(\nu) = \mathrm{E}\left\{\left|\widetilde{F}_{00}(\nu)\right|^2\right\}$. As the diffuse signal is spherically isotropic, the coherence matrix is given by [3, 22, 47]

$$\mathbf{\Gamma}_{\widetilde{\mathbf{f}}}(\nu) = \mathbf{I}_{N\times N}, \qquad (9.44)$$

where $\mathbf{I}_{N\times N}$ denotes the $N \times N$ identity matrix.

Our task then becomes the estimation of the variance of $\widetilde{F}_{00}(\nu)$. In [3], Braun et al. proposed an informed spatial filtering approach to dereverberation, where $\phi_{\widetilde{F}_{00}}(\nu)$ was estimated using the SDR, introduced in Sect. 5.2. Indeed, the SDR is given by

$$\mathrm{SDR}(\nu) = \frac{\phi_{\widetilde{X}_{00}}(\nu)}{\phi_{\widetilde{F}_{00}}(\nu)}, \tag{9.45}$$

allowing us to express $\phi_{\widetilde{F}_{00}}(\nu)$ as

$$\phi_{\widetilde{F}_{00}}(\nu) = \frac{\phi_{\widetilde{X}_{00}}(\nu)}{\mathrm{SDR}(\nu)} \tag{9.46a}$$

$$= \frac{\phi_{\widetilde{X}_{00}}(\nu)}{\mathrm{SDR}(\nu)} \frac{1+\mathrm{SDR}^{-1}(\nu)}{1+\mathrm{SDR}^{-1}(\nu)} \tag{9.46b}$$

$$= \frac{\phi_{\widetilde{X}_{00}}(\nu) + \phi_{\widetilde{F}_{00}}(\nu)}{\mathrm{SDR}(\nu)+1} \tag{9.46c}$$

$$= \frac{\phi_{\widetilde{P}_{00}}(\nu) - \phi_{\widetilde{V}_{00}}(\nu)}{\mathrm{SDR}(\nu)+1}, \tag{9.46d}$$

where $\phi_{\widetilde{P}_{00}}(\nu) = \mathrm{E}\left\{\left|\widetilde{P}_{00}(\nu)\right|^2\right\}$, $\phi_{\widetilde{V}_{00}}(\nu) = \mathrm{E}\left\{\left|\widetilde{V}_{00}(\nu)\right|^2\right\}$, and (9.46d) is obtained using the relation $\phi_{\widetilde{P}_{00}}(\nu) = \phi_{\widetilde{X}_{00}}(\nu) + \phi_{\widetilde{F}_{00}}(\nu) + \phi_{\widetilde{V}_{00}}(\nu)$, which holds because the three eigenbeams are mutually uncorrelated.

This informed spatial filtering approach to dereverberation is summarized in the form of a block diagram in Fig. 9.6. The spatial domain sound pressure signals $P(\nu, \mathbf{r}_q)$ are transformed to the SHD, and mode strength compensated using (9.29). The SDR is estimated (see Sect. 5.2), and the interference PSD matrix $\mathbf{\Phi}_{\widetilde{\mathbf{u}}}(\nu)$ is then estimated using (9.42), (9.43) and (9.46d). Finally, the multichannel Wiener filter weights are computed using Ω_{dir} and the interference PSD matrix $\mathbf{\Phi}_{\widetilde{\mathbf{u}}}(\nu)$, and the filter is applied to the mode strength compensated eigenbeams to yield the filter output $Z(\nu)$.

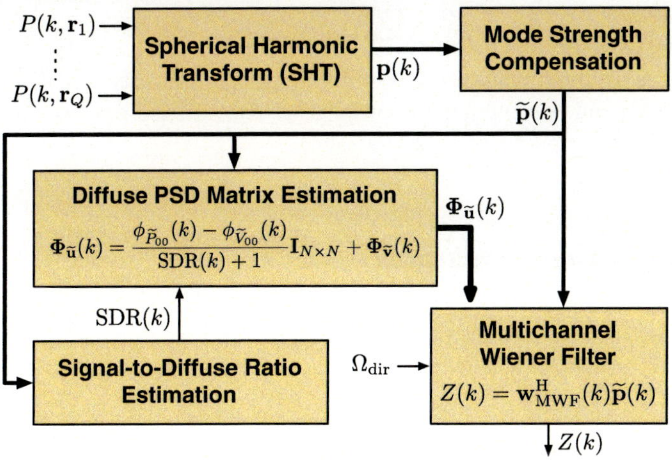

Fig. 9.6 Block diagram of the informed spatial filtering approach to dereverberation

9.2.3 Relation to Robust MVDR Filter

The *robust* MVDR filter [7] uses diagonal loading to improve the robustness of the MVDR filter to errors in microphone placement and steering direction. The weights of the robust MVDR filter are given by

$$\mathbf{w}_{\mathrm{rMVDR}}(k, \delta_{\mathrm{r}}) = \frac{\left[\boldsymbol{\Phi}_{\tilde{\mathbf{v}}}(\nu) + \delta_{\mathrm{r}}(\nu)\mathbf{I}_{N\times N}\right]^{-1}\mathbf{d}_{\mathrm{dir}}}{\mathbf{d}_{\mathrm{dir}}^{\mathrm{H}}\left[\boldsymbol{\Phi}_{\tilde{\mathbf{v}}}(\nu) + \delta_{\mathrm{r}}(\nu)\mathbf{I}_{N\times N}\right]^{-1}\mathbf{d}_{\mathrm{dir}}}, \tag{9.47}$$

where the regularization parameter $\delta_{\mathrm{r}}(\nu)$ simulates the presence of additional spatially white noise. Unlike the MVDR filter in (9.38), the robust MVDR filter does not depend on the variance $\phi_{\tilde{F}_{00}}(\nu)$, and the regularization parameter is usually time- and frequency-independent. If $\phi_{\tilde{F}_{00}}(\nu)$ is known and the regularization parameter is chosen as $\delta_{\mathrm{r}}(\nu) = \phi_{\tilde{F}_{00}}(\nu)$, the robust MVDR filter is equal to the MVDR filter in (9.38), that is, $\mathbf{w}_{\mathrm{rMVDR}}(k, \phi_{\tilde{F}_{00}}(\nu)) = \mathbf{w}_{\mathrm{MVDR}}(\nu)$.

When a fixed value of $\delta_{\mathrm{r}} = 0$ is chosen, the robust MVDR filter weights are given by

$$\mathbf{w}_{\mathrm{rMVDR}}(k, 0) = \frac{\boldsymbol{\Phi}_{\tilde{\mathbf{v}}}^{-1}(\nu)\mathbf{d}_{\mathrm{dir}}}{\mathbf{d}_{\mathrm{dir}}^{\mathrm{H}}\boldsymbol{\Phi}_{\tilde{\mathbf{v}}}^{-1}(\nu)\mathbf{d}_{\mathrm{dir}}}, \tag{9.48}$$

and the robust MVDR filter reduces to the MVDR filter in (9.38) with $\boldsymbol{\Phi}_{\tilde{\mathbf{f}}}(\nu) = \mathbf{0}_{N\times N}$, where $\mathbf{0}_{N\times N}$ denotes an $N \times N$ matrix of zeros. On the other hand, when δ_{r} tends to infinity, the robust MVDR filter weights are given by

$$\lim_{\delta\to\infty} \mathbf{w}_{\mathrm{rMVDR}}(k, \delta) = \frac{\mathbf{d}_{\mathrm{dir}}}{\mathbf{d}_{\mathrm{dir}}^{\mathrm{H}}\mathbf{d}_{\mathrm{dir}}}, \tag{9.49}$$

and the robust MVDR filter reduces to the MVDR filter in (9.38) with $\boldsymbol{\Phi}_{\tilde{\mathbf{v}}}(\nu) = \mathbf{0}_{N\times N}$. This filter is a maximum directivity beamformer, as per Property 7.1. Neither of the filters in (9.48) and (9.49) require knowledge of the variance $\phi_{\tilde{F}_{00}}(\nu)$; they will be used for comparison purposes in Sect. 9.2.4.

9.2.4 Performance Evaluation

In this section, we evaluate the performance of the informed spatial filter presented in Sect. 9.2.2, and compare it to the performance of the robust MVDR filter presented in Sect. 9.2.3.

Experimental Setup

We computed the sound pressure signals measured by a rigid spherical microphone array by simulating impulse responses with SMIRgen [17], an AIR simulator for spherical microphone arrays based on the algorithm presented in Chap. 4. The array was composed of $Q = 32$ microphones, and had a radius of 4.2 cm. The room dimensions were $5 \times 7 \times 4$ m and its reverberation time was 500 ms. The signals were processed in the STFT domain at a sampling frequency of 8 kHz, with a frame length of 32 ms, a 50 % overlap between successive frames, and a fast Fourier transform length of 64 ms in order to avoid circular convolution errors.

The source producing the direct and diffuse signals was placed at a distance of 1 m from the centre of the array, and its DOA was assumed to be known for the purposes of computing the steering vector $\mathbf{d}_{\text{dir}}$. The source signal consisted of 5 s of male speech from the EBU SQAM dataset [9]. The noise signal consisted of spatio-temporally white Gaussian noise with a signal-to-noise ratio (SNR) of 25 dB at the microphone closest to the source.

The beamformer was applied to eigenbeams of orders up to $L = 3$. The PSD matrices were estimated recursively with a time constant of 30 ms; the noise PSD matrix was computed using only the first 50 frames of the eigenbeam $\widetilde{P}_{00}$, where the direct and diffuse signals were not present. The SDR was estimated using the coefficient of variation (CV) method, as presented in Sect. 5.2.2, and the expectations in (5.58) were estimated using moving averages over 8 time frames.

Results

In order to illustrate the effect of the informed spatial filter, for one example input signal we begin by plotting in Fig. 9.7 spectrograms of four signals: (a) the desired, anechoic signal $\widetilde{X}_{00}(\nu)$; (b) the filter input signal including reverberation, $\widetilde{X}_{00}(\nu) + \widetilde{F}_{00}(\nu)$; (c) the noisy filter input signal including both sensor noise and reverberation, $\widetilde{P}_{00}(\nu) = \widetilde{X}_{00}(\nu) + \widetilde{F}_{00}(\nu) + \widetilde{V}_{00}(\nu)$; and (d) the filter output signal $Z(\nu)$. By comparing (a) and (b), it can be seen that reverberation causes a temporal smearing effect, blurring the boundaries between phonemes. In (d), the smearing is reduced, and some of the sensor noise has also been suppressed; the filter has performed both dereverberation and noise reduction, as desired.

The accurate estimation of the diffuse PSD $\phi_{\widetilde{F}_{00}}(\nu)$ is crucial to the performance of the informed spatial filter $\mathbf{w}_{\text{MWF}}$. In Fig. 9.8, we plot the ideal and estimated diffuse PSDs. We see that the estimated diffuse PSD faithfully tracks changes in the ideal diffuse PSD, and that the PSD values are also accurately estimated. The diffuse PSD is, however, slightly overestimated due to the presence of sensor noise.

We analyzed the narrowband performance of the robust MVDR filter $\mathbf{w}_{\text{rMVDR}}$ and the MVDR filter $\mathbf{w}_{\text{MVDR}}$ that forms part of the informed spatial filter $\mathbf{w}_{\text{MWF}}$ in (9.38), using two performance measures: the noise reduction factor (NRF) and the directivity index (DI). The NRF was defined in the same way as in Chap. 7, and is given by

$$\xi_{\text{nr}}\left[\mathbf{w}(\nu)\right] = \frac{\phi_{\widetilde{V}_{00}}(\nu)}{\mathbf{w}^{\text{H}}(\nu)\boldsymbol{\Phi}_{\widetilde{\mathbf{v}}}(\nu)\mathbf{w}(\nu)}. \tag{9.50}$$

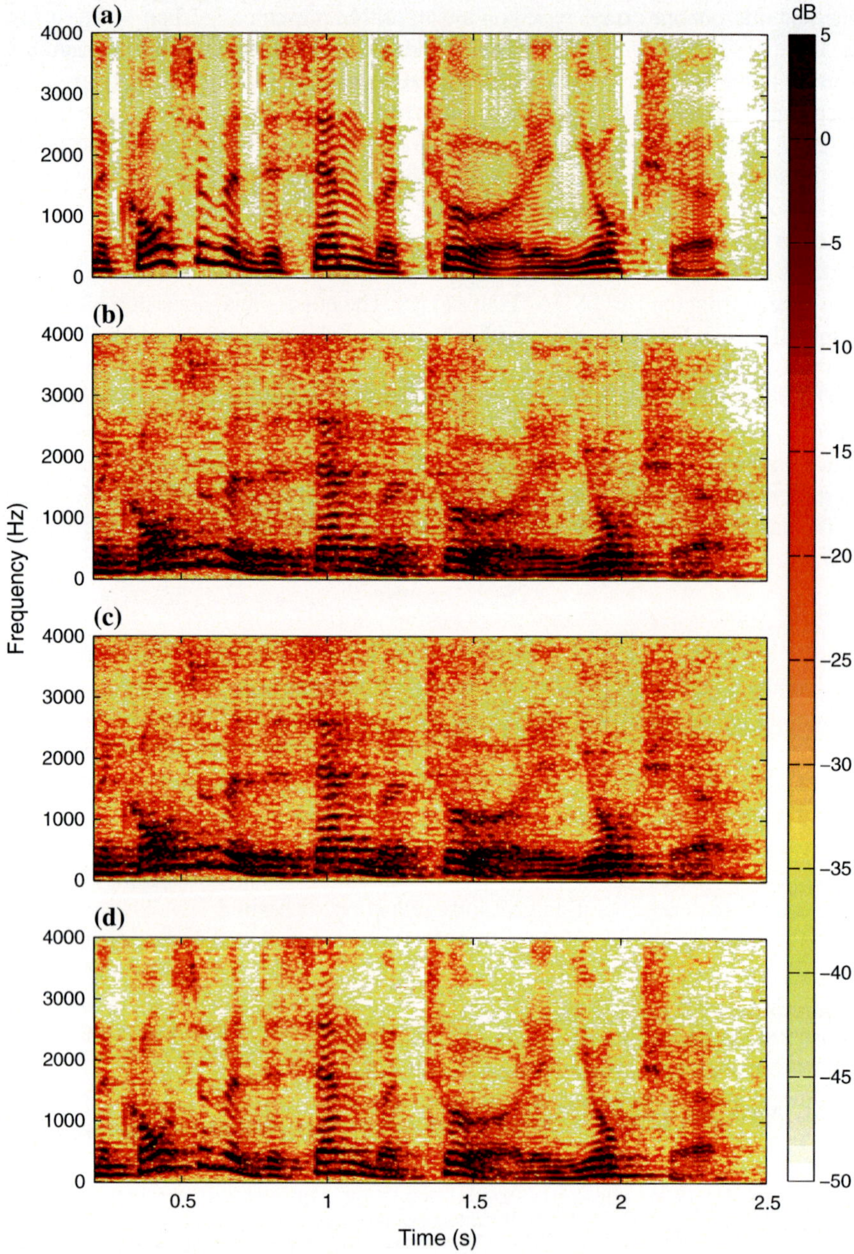

Fig. 9.7 Sample spectrograms of **a** anechoic desired signal $\widetilde{X}_{00}(\nu)$, **b** reverberant filter input signal $\widetilde{X}_{00}(\nu) + \widetilde{F}_{00}(\nu)$, **c** noisy and reverberant filter input signal $\widetilde{P}_{00}(\nu) = \widetilde{X}_{00}(\nu) + \widetilde{F}_{00}(\nu) + \widetilde{V}_{00}(\nu)$ and **d** filter output signal $Z(\nu)$

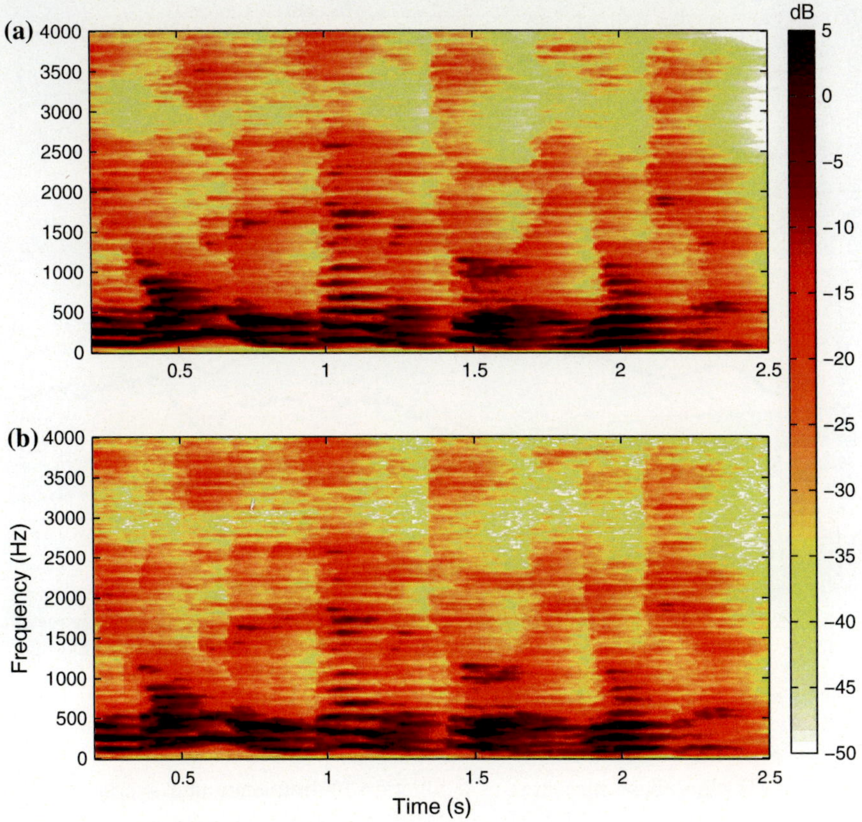

Fig. 9.8 **a** Ideal and **b** estimated diffuse PSD $\phi_{\widetilde{F}_{00}}(\nu)$

The DI measure we used is similar to the DI defined in Chap. 6, and is given by

$$\mathrm{DI}(\nu) = \frac{|\mathbf{w}^{\mathrm{H}}(\nu)\mathbf{d}_{\mathrm{dir}}(\nu)|^2}{\mathbf{w}^{\mathrm{H}}(\nu)\boldsymbol{\Gamma}_{\widetilde{\mathbf{f}}}(\nu)\mathbf{w}(\nu)}. \tag{9.51}$$

Since the MVDR and robust MVDR filters satisfy the distortionless constraint $\mathbf{w}^{\mathrm{H}}(\nu)\mathbf{d}_{\mathrm{dir}}(\nu) = 1$, and the coherence matrix $\boldsymbol{\Gamma}_{\widetilde{\mathbf{f}}}(\nu)$ is an identity matrix, this can be simplified to[7]

$$\mathrm{DI}(\nu) = \frac{1}{\mathbf{w}^{\mathrm{H}}(\nu)\mathbf{w}(\nu)}. \tag{9.52}$$

[7] It should be noted that this simplified expression is only valid if the filter is applied to mode strength compensated eigenbeams. As a result, it is different to the expression given in Chap. 6.

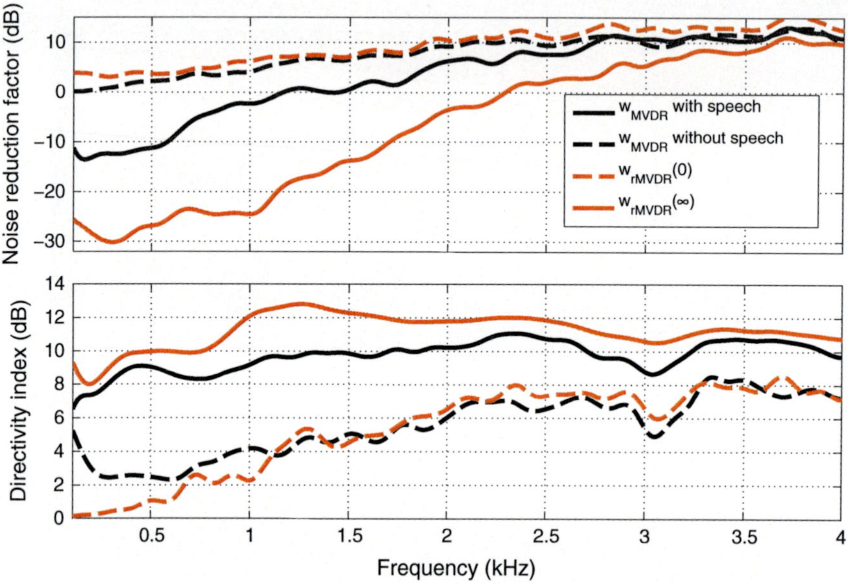

Fig. 9.9 Average NRF and DI for three different filters: the MVDR filter $\mathbf{w}_{\mathrm{MVDR}}$ in the presence and absence of speech, the robust MVDR filter $\mathbf{w}_{\mathrm{rMVDR}}(0)$ for $\delta_{\mathrm{r}} = 0$, and the robust MVDR filter $\mathbf{w}_{\mathrm{rMVDR}}(\infty)$ for $\delta_{\mathrm{r}} \to \infty$. When $\delta = 0$, the robust MVDR filter maximizes the NRF; when $\delta \to \infty$, the robust MVDR filter maximizes the DI

Figure 9.9 plots these measures as a function of frequency and averaged over time. In the case of the MVDR filter, whose weights depend on the interference PSD matrix $\mathbf{\Phi}_{\tilde{\mathbf{u}}}(\nu)$, we computed the averages separately over time frames where speech is present and where speech is absent. As expected, the robust MVDR filters $\mathbf{w}_{\mathrm{rMVDR}}(k, 0)$ and $\mathbf{w}_{\mathrm{rMVDR}}(k, \infty)$ set bounds for the performance of the MVDR filter. In the absence of speech, the MVDR filter converges to the robust MVDR filter $\mathbf{w}_{\mathrm{rMVDR}}(k, 0)$, achieving the highest NRF and the lowest DI. In the presence of speech, the MVDR filter converges to the robust MVDR filter $\mathbf{w}_{\mathrm{rMVDR}}(k, \infty)$, achieving the lowest NRF and the highest DI. At high frequencies, the speech has low energy, and the performance of the MVDR filter is therefore similar whether speech is present or not. At some frequencies and for some filters, the NRF is below 1 (or equivalently, below 0 dB), indicating that the power of the noise at the output of the filter is higher than at the reference microphone $\mathcal{M}_{\mathrm{ref}}$. Nevertheless, as noted in Sect. 7.2.2, the power of spatially incoherent noise (as used in this experimental setup) is lower at $\mathcal{M}_{\mathrm{ref}}$ than at the individual microphones, because the reference microphone signal $\widetilde{P}_{00}(k)$ is formed using all Q individual microphone signals, and spatially incoherent noise sums destructively.

Finally, in Table 9.1, we present a number of broadband performance measures. The improvement in the segmental signal-to-interference ratio between the reference microphone $\mathcal{M}_{\mathrm{ref}}$ and the filter output is denoted as ΔsegSIR. The segmental DI and

Table 9.1 Performance of the informed spatial filter and the robust MVDR filters with and without the diffuse PSD estimator

	Wiener filters		Robust MVDR filters		
	$\mathbf{w}_{\mathrm{MWF}}$	$\mathbf{w}_{\mathrm{MWF,\,V}}$	$\mathbf{w}_{\mathrm{rMVDR}}(\phi_{\widetilde{F}_{00}})$	$\mathbf{w}_{\mathrm{rMVDR}}(0)$	$\mathbf{w}_{\mathrm{rMVDR}}(\infty)$
ΔsegSIR (dB)	**9.2**	6.8	**8.8**	6.7	7.7
segNRF (dB)	**10.6**	10.3	7.4	**9.4**	4.6
segDI (dB)	**13.4**	9.3	10.3	8.4	**10.9**
SRMR (dB)	**6.0**	3.2	5.6	3.2	**5.8**

segmental NRF were respectively computed using (9.51) and (9.50), and are denoted as segDI and segNRF. All segmental measures were computed over frequencies from 100 to 4 kHz. The speech-to-reverberation modulation energy ratio (SRMR), proposed in Falk et al. [10], is a non-intrusive quality and intelligibility measure for reverberant speech; low values of the SRMR are obtained when the speech is highly reverberant. In order to ensure that the SRMR only evaluated the dereverberation performance of the filters, it was computed based on noise-free versions of the filter output signals.

In the first column of Table 9.1, we see that the informed spatial filter $\mathbf{w}_{\mathrm{MWF}}$ has the best performance across all measures. In the second column, we show the performance of the informed spatial filter when it only seeks to suppress sensor noise, and does not seek to suppress the diffuse signal; this is achieved by setting $\mathbf{\Phi}_{\widetilde{\mathbf{f}}}(\nu) = 0$. The next three columns show the performance of the MVDR filter $\mathbf{w}_{\mathrm{MVDR}}$ in (9.38), which is equivalent to the robust MVDR filter with $\delta_{\mathrm{r}}(\nu) = \phi_{\widetilde{F}_{00}}(\nu)$; the robust MVDR filter with $\delta_{\mathrm{r}} = 0$; and the robust MVDR filter with $\delta_{\mathrm{r}} \to \infty$. The performance of the MVDR filters is consistent with Fig. 9.9: $\delta_{\mathrm{r}} = 0$ yields the best noise reduction performance and worst dereverberation performance; while $\delta_{\mathrm{r}} \to \infty$ yields the worst noise reduction performance and best dereverberation performance. The choice $\delta_{\mathrm{r}}(\nu) = \phi_{\widetilde{F}_{00}}(\nu)$ achieves a good tradeoff between noise reduction and dereverberation, yielding the highest segSIR improvement of all the robust MVDR filters. These results are in line with those of informal listening tests.[8] The SRMR values obtained at the output of the filters can be compared to the following reference values, in order to evaluate the absolute dereverberation performance: the SRMR of the signal $\widetilde{P}_{00}(\nu)$ at the reference microphone $\mathcal{M}_{\mathrm{ref}}$ was 2.9, and the SRMR of the desired signal $\widetilde{X}_{00}(\nu)$ was 9.1.

[8] A number of audio examples can be accessed from https://www.audiolabs-erlangen.de/resources/2013-ICASSP-RR.

9.3 Chapter Summary and Conclusions

This chapter provided an introduction to the informed array processing approach and showed its application to two problems: coherent and incoherent noise reduction, and joint dereverberation and incoherent noise reduction. In the first problem, instantaneous narrowband DOA estimates were employed in order to distinguish between desired and undesired spatially coherent sources. The DOA estimates were used to estimate a DSPP, which was in turn used to estimate the second-order statistics of the desired speech and noise. Finally, the estimated statistics were applied to a tradeoff beamformer. It was shown that the informed noise reduction algorithm is able to suppress high levels of coherent noise.

In the second problem, instantaneous narrowband SDR estimates were employed to suppress diffuse sound. The SDR estimates were used to estimate the second-order statistics of the diffuse sound, which were combined with the DOA of the desired source to compute the weights of a multichannel Wiener filter. It was shown that the informed dereverberation method achieves an optimal tradeoff between noise reduction and dereverberation, with high values of the segmental NRF and segmental DI.

The estimation of the second-order statistics in signal enhancement problems remains a challenge and a topic of active research, both in the spatial domain and in the SHD. Future research aims include the estimation of these statistics for noise reduction purposes, in the presence of one or more diffuse sources and/or multiple desired coherent sources, and for dereverberation purposes, in the presence of multiple coherent sources.

References

1. Berge, S., Barrett, N.: High angular resolution planewave expansion. In: Proceedings of the 2nd International Symposium on Ambisonics and Spherical Acoustics (2010)
2. Bradley, J.S., Sato, H., Picard, M.: On the importance of early reflections for speech in rooms. J. Acoust. Soc. Am. **113**(6), 3233–3244 (2003)
3. Braun, S., Jarrett, D.P., Fischer, J., Habets, E.A.P.: An informed spatial filter for dereverberation in the spherical harmonic domain. In: Proceedings of the IEEE International Conference on Acoustics, Speech and Signal Processing (ICASSP), pp. 669–673. Vancouver, Canada (2013)
4. Cohen, I.: Noise spectrum estimation in adverse environments: Improved minima controlled recursive averaging. IEEE Trans. Speech Audio Process. **11**(5), 466–475 (2003). doi:10.1109/TSA.2003.811544
5. Cohen, I.: Multichannel post-filtering in nonstationary noise environments. IEEE Trans. Signal Process. **52**(5), 1149–1160 (2004)
6. Cohen, I., Gannot, S., Berdugo, B.: An integrated real-time beamforming and post filtering system for nonstationary noise environments. EURASIP J. Appl. Signal Process. **11**, 1064–1073 (2003)
7. Cox, H., Zeskind, R.M., Owen, M.M.: Robust adaptive beamforming. IEEE Trans. Acoust., Speech, Signal Process. **35**(10), 1365–1376 (1987)

8. Ephraim, Y., Malah, D.: Speech enhancement using a minimum-mean square error short-time spectral amplitude estimator. IEEE Trans. Acoust., Speech, Signal Process. **32**(6), 1109–1121 (1984)
9. European Broadcasting Union: Sound quality assessment material recordings for subjective tests. http://tech.ebu.ch/publications/sqamcd (1988)
10. Falk, T., Zheng, C., Chan, W.Y.: A non-intrusive quality and intelligibility measure of reverberant and dereverberated speech. IEEE Trans. Audio, Speech, Lang. Process. **18**(7), 1766–1774 (2010)
11. Fisher, R.: Dispersion on a sphere. Proc. R. Soc. Lond. Ser. A **217**(1130), 295–305 (1953). doi:10.1098/rspa.1953.0064
12. Habets, E.A.P.: Single- and multi-microphone speech dereverberation using spectral enhancement. Ph.D. thesis, Technische Universiteit Eindhoven. http://alexandria.tue.nl/extra2/200710970.pdf (2007)
13. Habets, E.A.P.: A distortionless subband beamformer for noise reduction in reverberant environments. In: Proceedings of the International Workshop on Acoustic Signal Enhancement (IWAENC), pp. 1–4 (2010)
14. Habets, E.A.P., Benesty, J., Cohen, I., Gannot, S., Dmochowski, J.: New insights into the MVDR beamformer in room acoustics. IEEE Trans. Audio, Speech, Lang. Process. **18**, 158–170 (2010)
15. Hendriks, R., Gerkmann, T.: Noise correlation matrix estimation for multi-microphone speech enhancement. IEEE Trans. Audio, Speech, Lang. Process. **20**(1), 223–233 (2012)
16. ITU-T: Objective measurement of active speech level (1993)
17. Jarrett, D.P.: Spherical microphone array impulse response (SMIR) generator. http://www.ee.ic.ac.uk/sap/smirgen/
18. Jarrett, D.P., Habets, E.A.P.: On the noise reduction performance of a spherical harmonic domain tradeoff beamformer. IEEE Signal Process. Lett. **19**(11), 773–776 (2012)
19. Jarrett, D.P., Habets, E.A.P., Naylor, P.A.: 3D source localization in the spherical harmonic domain using a pseudointensity vector. In: Proceedings of the European Signal Processing Conference (EUSIPCO), pp. 442–446. Aalborg, Denmark (2010)
20. Jarrett, D.P., Habets, E.A.P., Thomas, M.R.P., Gaubitch, N.D., Naylor, P.A.: Dereverberation performance of rigid and open spherical microphone arrays: theory & simulation. In: Proceedings of the Joint Workshop on Hands-Free Speech Communication and Microphone Arrays (HSCMA), pp. 145–150. Edinburgh, UK (2011)
21. Jarrett, D.P., Habets, E.A.P., Benesty, J., Naylor, P.A.: A tradeoff beamformer for noise reduction in the spherical harmonic domain. In: Proceedings of the International Workshop on Acoustic Signal Enhancement (IWAENC). Aachen, Germany (2012)
22. Jarrett, D.P., Thiergart, O., Habets, E.A.P., Naylor, P.A.: Coherence-based diffuseness estimation in the spherical harmonic domain. In: Proceedings of the IEEE Convention of Electrical & Electronics Engineers in Israel (IEEEI). Eilat, Israel (2012)
23. Jarrett, D.P., Habets, E.A.P., Naylor, P.A.: Spherical harmonic domain noise reduction using an MVDR beamformer and DOA-based second-order statistics estimation. In: Proceedings of the IEEE International Conference on Acoustics, Speech and Signal Processing (ICASSP), pp. 654–658. Vancouver, Canada (2013)
24. Jarrett, D.P., Taseska, M., Habets, E.A.P., Naylor, P.A.: Noise reduction in the spherical harmonic domain using a tradeoff beamformer and narrowband DOA estimates. IEEE/ACM Trans. Audio, Speech, Lang. Process. **22**(5), 965–976 (2014)
25. Kuttruff, H.: Room Acoustics, 4th edn. Taylor & Francis, London (2000)
26. Mardia, K.V., Jupp, P.E.: Directional Statistics. Wiley-Blackwell, New York (1999)
27. McCowan, I., Bourlard, H.: Microphone array post-filter based on noise field coherence. IEEE Trans. Speech Audio Process. **11**(6), 709–716 (2003)
28. Meyer, J., Elko, G.: A highly scalable spherical microphone array based on an orthonormal decomposition of the soundfield. In: Proceedings of the IEEE International Conference on Acoustics, Speech and Signal Processing (ICASSP), vol. 2, pp. 1781–1784 (2002)
29. Nábělek, A.K., Mason, D.: Effect of noise and reverberation on binaural and monaural word identification by subjects with various audiograms. J Speech Hear. Res. **24**, 375–383 (1981)

30. Naylor, P.A., Gaubitch, N.D. (eds.): Speech Dereverberation. Springer, Heidelberg (2010)
31. Ngo, K., Spriet, A., Moonen, M., Wouters, J., Jensen, S.: Incorporating the conditional speech presence probability in multi-channel Wiener filter based noise reduction in hearing aids. EURASIP J. Adv. Signal Process. (1) (2009). doi:10.1155/2009/930625 (Special Issue on Digital Signal Processing for Hearing Instruments)
32. Peled, Y., Rafaely, B.: Study of speech intelligibility in noisy enclosures using spherical microphones arrays. In: Proceedings of the Joint Workshop on Hands-Free Speech Communication and Microphone Arrays (HSCMA), pp. 160–163 (2008). doi:10.1109/HSCMA.2008.4538711
33. Peled, Y., Rafaely, B.: Method for dereverberation and noise reduction using spherical microphone arrays. In: Proceedings of the IEEE International Conference on Acoustics, Speech and Signal Processing (ICASSP), pp. 113–116 (2010). doi:10.1109/ICASSP.2010.5496154
34. Peled, Y., Rafaely, B.: Linearly constrained minimum variance method for spherical microphone arrays in a coherent environment. In: Proceedings of the Hands-Free Speech Communication and Microphone Arrays (HSCMA), pp. 86–91 (2011). doi:10.1109/HSCMA.2011.5942416
35. Pulkki, V.: Spatial sound reproduction with directional audio coding. J. Audio Eng. Soc. **55**(6), 503–516 (2007)
36. Rafaely, B.: Plane-wave decomposition of the pressure on a sphere by spherical convolution. J. Acoust. Soc. Am. **116**(4), 2149–2157 (2004)
37. Rickard, S., Yilmaz, Z.: On the approximate W-disjoint orthogonality of speech. In: Proceedings of the IEEE International Conference on Acoustics, Speech and Signal Processing (ICASSP), vol. 1, pp. 529–532 (2002)
38. Silzle, A., Geyersberger, S., Brohasga, G., Weninger, D., Leistner, M.: Vision and technique behind the new studios and listening rooms of the Fraunhofer IIS audio laboratory. In: Proceedings of the Audio Engineering Society Convention (2009)
39. Souden, M., Chen, J., Benesty, J., Affes, S.: Gaussian model-based multichannel speech presence probability. IEEE Trans. Audio, Speech, Lang. Process. **18**(5), 1072–1077 (2010)
40. Souden, M., Chen, J., Benesty, J., Affes, S.: An integrated solution for online multichannel noise tracking and reduction. IEEE Trans. Audio, Speech, Lang. Process. **19**(7), 2159–2169 (2011). doi:10.1109/TASL.2011.2118205
41. Sra, S.: A short note on parameter approximation for von Mises-Fisher distributions: and a fast implementation of $I_s(x)$. Comput. Stat. **27**(1), 177–190 (2012). doi:10.1007/s00180-011-0232-x
42. Steinberg, J.C.: Effects of distortion upon the recognition of speech sounds. J. Acoust. Soc. Am. **1**, 35–35 (1929)
43. Taseska, M., Habets, E.A.P.: MMSE-based blind source extraction in diffuse noise fields using a complex coherence-based a priori SAP estimator. In: Proceedings of the International Workshop on Acoustic Signal Enhancement (IWAENC) (2012)
44. Teutsch, H.: Wavefield decomposition using microphone arrays and its application to acoustic scene analysis. Ph.D. thesis, Friedrich-Alexander Universität Erlangen-Nürnberg (2005)
45. Viola, P., Jones, M.J.: Rapid object detection using a boosted cascade of simple features. In: Proceedings of the IEEE Computer Society Conference on Computer Vision and Pattern Recognition, vol. 1, pp. 511–518 (2001). doi:10.1109/CVPR.2001.990517
46. Wu, P.K.T., Epain, N., Jin, C.: A dereverberation algorithm for spherical microphone arrays using compressed sensing techniques. In: Proceedings of the IEEE International Conference on Acoustics, Speech and Signal Processing (ICASSP), pp. 4053–4056 (2012)
47. Yan, S., Sun, H., Svensson, U.P., Ma, X., Hovem, J.M.: Optimal modal beamforming for spherical microphone arrays. IEEE Trans. Audio, Speech, Lang. Process. **19**(2), 361–371 (2011). doi:10.1109/TASL.2010.2047815

Index

D. Jarrett et al., *Theory and Applications of Spherical Microphone Array Processing*, Springer Topics in Signal Processing 9,
DOI 10.1007/978-3-319-42211-4

Printed in the United States
By Bookmasters